International Political Economy Series

General Editor: **Timothy M. Shaw**, Professor of Political Science and International Development Studies, and Director of the Centre for Foreign Policy Studies, Dalhousie University, Halifax, Nova Scotia

Recent titles include:

Pradeep Agrawal, Subir V. Gokarn, Veena Mishra, Kirit S. Parikh and Kunal Sen
ECONOMIC RESTRUCTURING IN EAST ASIA AND INDIA
Perspectives on Policy Reform

Deborah Bräutigam
CHINESE AID AND AFRICAN DEVELOPMENT
Exporting Green Revolution

Steve Chan, Cal Clark and Danny Lam (*editors*)
BEYOND THE DEVELOPMENTAL STATE
East Asia's Political Economies Reconsidered

Jennifer Clapp
ADJUSTMENT AND AGRICULTURE IN AFRICA
Farmers, the State and the World Bank in Guinea

Robert W. Cox (*editor*)
THE NEW REALISM
Perspectives on Multilateralism and World Order

Ann Denholm Crosby
DILEMMAS IN DEFENCE DECISION-MAKING
Constructing Canada's Role in NORAD, 1958–96

Diane Ethier
ECONOMIC ADJUSTMENT IN NEW DEMOCRACIES
Lessons from Southern Europe

Stephen Gill (*editor*)
GLOBALIZATION, DEMOCRATIZATION AND MULTILATERALISM

Jeffrey Henderson (*editor*), assisted by Karoly Balaton and Gyorgy Lengyel
INDUSTRIAL TRANSFORMATION IN EASTERN EUROPE IN THE LIGHT OF THE EAST ASIAN EXPERIENCE

Jacques Hersh and Johannes Dragsbaek Schmidt (*editors*)
THE AFTERMATH OF 'REAL EXISTING SOCIALISM' IN EASTERN EUROPE,
Volume 1: Between Western Europe and East Asia

David Hulme and Michael Edwards (*editors*)
NGOs, STATES AND DONORS: Too Close for Comfort?

Staffan Lindberg and Árni Sverrisson (*editors*)
SOCIAL MOVEMENTS IN DEVELOPMENT
The Challenge of Globalization and Democratization

Anne Lorentzen and Marianne Rostgaard (*editors*)
THE AFTERMATH OF 'REAL EXISTING SOCIALISM' IN EASTERN EUROPE,
Volume 2: People and Technology in the Process of Transition

Stephen D. McDowell
GLOBALIZATION, LIBERALIZATION AND POLICY CHANGE
A Political Economy of India's Communications Sector

Juan Antonio Morales and Gary McMahon (*editors*)
ECONOMIC POLICY AND THE TRANSITION TO DEMOCRACY
The Latin American Experience

Ted Schrecker (*editor*)
SURVIVING GLOBALISM
The Social and Environmental Challenges

Ann Seidman, Robert B. Seidman and Janice Payne (*editors*)
LEGISLATIVE DRAFTING FOR MARKET REFORM
Some Lessons from China

Caroline Thomas and Peter Wilkin (*editors*)
GLOBALIZATION AND THE SOUTH

Kenneth P. Thomas
CAPITAL BEYOND BORDERS
States and Firms in the Auto Industry, 1960–94

Geoffrey R. D. Underhill (*editors*)
THE NEW WORLD ORDER IN INTERNATIONAL FINANCE

Henry Veltmeyer, James Petras and Steve Vieux
NEOLIBERALISM AND CLASS CONFLICT IN LATIN AMERICA
A Comparative Perspective on the Political Economy of Structural Adjustment

Robert Wolfe
FARM WARS
The Political Economy of Agriculture and the International Trade Regime

---

International Political Economy Series
Series Standing Order ISBN 0–333–71708–2 hardcover
Series Standing Order ISBN 0–333–71110–6 paperback
(*outside North America only*)

You can receive future titles in this series as they are published by placing a standing order. Please contact your bookseller or, in case of difficulty, write to us at the address below with your name and address, the title of the series and one of the ISBNs quoted above.

Customer Services Department, Macmillan Distribution Ltd, Houndmills, Basingstoke, Hampshire RG21 6XS, England

Environment and Security

Discourses and Practices

Edited by

Miriam R. Lowi
Associate Professor of Political Science
The College of New Jersey
Ewing, New Jersey
USA

and

Brian R. Shaw
Manager, Center for Environmental Security
Pacific Northwest National Laboratory
Washington, DC
USA

First published in Great Britain 2000 by
MACMILLAN PRESS LTD
Houndmills, Basingstoke, Hampshire RG21 6XS and London
Companies and representatives throughout the world

A catalogue record for this book is available from the British Library.

ISBN 0–333–73167–0

 First published in the United States of America 2000 by
ST. MARTIN'S PRESS, INC.,
Scholarly and Reference Division,
175 Fifth Avenue, New York, N.Y. 10010

ISBN 0–312–22485–0

Library of Congress Cataloging-in-Publication Data
Environment and security : discourses and practices / edited by Miriam R. Lowi and Brian R. Shaw.
p. cm. — (International political economy series)
Includes bibliographical references and index.
ISBN 0–312–22485–0 (cloth)
1. Environmental policy. 2. National security—Environmental aspects. I. Lowi, Miriam R. II. Shaw, Brian R., 1950– .
III. Series.
GE170.E574 1999
363.7'056—dc21   99–33852
                                                CIP

Selection, editorial matter and Chapter 1 © Miriam R. Lowi and Brian R. Shaw 2000
Chapter 9 © Miriam R. Lowi 2000
Chapters 2–8, 10, 11 © Macmillan Press Ltd 2000

All rights reserved. No reproduction, copy or transmission of this publication may be made without written permission.

No paragraph of this publication may be reproduced, copied or transmitted save with written permission or in accordance with the provisions of the Copyright, Designs and Patents Act 1988, or under the terms of any licence permitting limited copying issued by the Copyright Licensing Agency, 90 Tottenham Court Road, London W1P 0LP.

Any person who does any unauthorised act in relation to this publication may be liable to criminal prosecution and civil claims for damages.

The authors have asserted their rights to be identified as the authors of this work in accordance with the Copyright, Designs and Patents Act 1988.

This book is printed on paper suitable for recycling and made from fully managed and sustained forest sources.

10 9 8 7 6 5 4 3 2 1
09 08 07 06 05 04 03 02 01 00

Printed and bound in Great Britain by
Antony Rowe Ltd, Chippenham, Wiltshire

*For Jazia and Ismael,
Brian and George*

Contents

List of Figures and Map	ix
List of Tables	x
Acknowledgements	xi
List of Acronyms	xii
Notes on the Contributors	xiii

1 Introduction and Overview
  Miriam R. Lowi and Brian R. Shaw … 1

Part I: Discourses … 9

2 The Changing Definition of National Security
  Mary Margaret Evans, John W. Mentz, Robert W. Chandler, and Stephanie L. Eubanks … 11

3 Integrating Environmental Factors into Conventional Security
  Richard A. Matthew … 33

4 Security, Governance and the Environment
  Steve Rayner and Elizabeth L. Malone … 49

5 Human Security, Environmental Security and Sustainable Development
  Steve Lonergan … 66

6 Geopolitics and Ecology: Rethinking the Contexts of Environmental Security
  Simon Dalby … 84

Part II: Practices … 101

7 Integration of Non-Traditional Security Issues: a Preliminary Application to South Korea
  Robert E. Bedeski … 103

8 Trends in Transboundary Water Disputes and Dispute Resolution
   Aaron T. Wolf and Jesse H. Hamner 123

9 Water and Conflict in the Middle East and South Asia
   Miriam R. Lowi 149

10 Perceptions of Risk and Security: the Aral Sea Basin
   Galina Sergen and Elizabeth L. Malone 172

11 Not Seeing the People for the Population: a Cautionary Tale from the Himalaya
   Michael Thompson 192

Bibliography 207

Index 222

List of Figures and Map

3.1	Causal model of environmental change and security	40
3.2	Development trap	41
3.3	Motivation/capability grid	42
4.1	Actual effects among civic involvement, socioeconomic development, and institutional performance: Italy, 1900s–1980s	55
4.2	Two-dimensional space is required for complex strategy switching	62
5.1	Conceptual diagram of environment and human security linkages	76
11.1	The Himalaya and their immediate surroundings	193

List of Tables

3.1	Perspectives on environment and security	36
3.2	Social responses to environmental change	39
4.1	Putnam's index of institutional performance	53
4.2	Characteristics of three kinds of social solidarity	61
4.3	Three institutional strategies	63
8.1	Shortened treaty titles, listed chronologically	131
8.2	Treaty statistics summary sheet	138

Acknowledgements

Most of the chapters in this volume were prepared for a multidisciplinary panel at the annual meeting of the American Association for the Advancement of Science, held in Seattle, Washington, in January 1997. The panel's aim was to integrate various approaches to understanding environmental security. Work leading to this volume was sponsored, in part, by the US Department of Energy, Office of Policy and International Affairs and the Office of Nonproliferation and National Security. The editors are grateful for the support provided by Norm Kahn, J. Kenneth Schafer, and Michael Ortmeier. The views and analyses expressed by the authors in this volume are their own, and do not necessarily reflect those of the agencies or institutions where they work.

Miriam R. Lowi put together this volume. She wrote the Introduction and Overview, in addition to her own Chapter 9; she edited the entire manuscript, and corrected the proofs. She worked directly with the staff at Macmillan, from the initial stages when she first approached the Press with the proposal for a collaborative work on environmental security, through the final, publication stages. Brian R. Shaw organized the panel that brought together seven of the contributors to this volume, and arranged funding for supporting resources and funding for secretarial support. He suggested a few revisions to the introductory chapter and prepared the index.

The editors would like to thank all contributors to this volume, and especially Richard A. Matthew and Galina Sergen for their assistance and valuable insights throughout. Joanne Linckhorst and Allyson Rucinski, of the College of New Jersey, provided excellent secretarial support, while the Department of Political Science most graciously provided logistical and general administrative support. Special thanks are due to Tim Shaw, the general editor of the International Political Economy Series, and to Aruna Vasudevan, the Commissioning Editor at Macmillan, for their patience and their skillful overseeing of the production of this volume through its various stages.

In Chapter 8, two passages are drawn from the authors' previous work: Hamner and Wolf (1998), published in *The Colorado Journal of International Environmental Law and Policy*, 1997 Yearbook; and Wolf (1997), published in *The International Journal of Water Resources Development*, 13:3, December. A version of Chapter 9 appeared in Lowi (1999), *The Journal of Environment and Development*, 8:4 (December), and is reproduced here by kind permission of the editor.

List of Acronyms

ABM	Anti-Ballistic Missile Treaty
ACI&C	Arms Control, Implementation, and Compliance
CFE	Conventional Armed Forces in Europe Treaty
CIA	Central Intelligence Agency, Government of the United States
CIDA	Canadian International Development Agency
DIA	Defense Intelligence Agency, Government of the United States
DOD	Department of Defense, Government of the United States
DPRK	Democratic People's Republic of Korea
GDP	Gross Domestic Product
IHDP/GECHS	International Human Dimensions Project of Global Environmental Change and Human Security
INF	Intermediate-Range Nuclear Forces Treaty
KEDO	Korean Peninsula Development Organisation
MAD	Mutually Assured Destruction
NATO	North Atlantic Treaty Organization
NIS	New Israeli Sheckels
NPCSD	North Pacific Cooperative Security Dialogue
NPT	Nuclear Non-Proliferation Treaty
ODA	Overseas Development Assistance
OPEC	Organization of Petroleum-Exporting Countries
PRC	People's Republic of China
START	Strategic Arms Reduction Treaty
TEK	Traditional Ecological Knowledge
TVA	Tennessee Valley Authority
UN	United Nations
UNCED	United Nations Conference on Environment and Development
UNDP	United Nations Development Programme
UNEP	United Nations Environment Programme
UNHCR	United Nations High Commissioner for Refugees
USAID	United States Agency for International Development
WCED	World Commission on Environment and Development
WRI	World Resources Institute

Notes on the Contributors

Robert E. Bedeski is a Professor of Political Science at the University of Victoria, British Columbia, Canada. He has served military tours in Korea and Taiwan, where he studied Chinese language for two years. He was a Japan Foundation Research Fellow twice, has lectured at numerous Asian universities, and was a Korea Foundation Research Fellow. He has written books and numerous articles on China, Japan, Korea, and is currently writing a book on human security, freedom and the Asian state. He serves as consultant to the Departments of Foreign Affairs and International Trade and of National Defense (Canada) in the area of Arms Control and Verification, and Non-Traditional Security Challenges. He is currently Co-Chair of the Canadian Consortium on Asia Pacific Security.

Robert W. Chandler is a Senior Councilor at the Atlantic Council of the United States and a member of the Council's NATO Advisory Committee. He writes in the areas of defense policy, military strategy, and combating terrorism.

Simon Dalby is an Associate Professor in the Department of Geography and Environmental Studies at Carleton University in Ottawa. His current research interests are in critical geopolitics and environmental security. Recently he has coedited *The Geopolitics Reader* and *Rethinking Geopolitics*, both published in 1998.

Stephanie L. Eubanks is an arms control policy analyst for the Office of the Under-Secretary of Defense, Office of Arms Control Implementation and Compliance (AC&C), Government of the United States.

Mary Margaret Evans is Deputy Director of the Arms Control, Implementation, and Compliance (ACI&C) Office in the Office of the Secretary of Defense, Government of the United States. As such, Ms Evans is responsible for Department of Defense planning and execution related to implementation of, and compliance with, all strategic nuclear and conventional arms control treaties and agreements. Before working in ACI&C, Ms Evans supported the Defense and Space Talks on behalf of the Strategic Defense Initiative Organization. She was the organization's representative to the Interagency Working Group from 1985 to 1987, and the Technical Advisor to the US Defense and Space Delegation in Geneva, Switzerland in 1988 and 1989.

Jesse H. Hamner is a doctoral student at Emory University in the Department of Political Science. His research focus is on political institutions that contribute to lasting agreements, primarily on natural resources. He is the co-creator of the Transboundary Freshwater Dispute Database.

Steve Lonergan is a Professor in the Department of Geography at the University of Victoria, British Columbia. He specializes in water issues in the Middle East, environment and security, and economic/ecological modelling. At the University of Victoria, he founded the Centre for Sustainable Regional Development and presently directs an international project on Global Environmental Change and Human Security for the International Human Dimensions of Global Change Programme, located in Bonn, Germany.

Miriam R. Lowi is Associate Professor in the Department of Political Science at the College of New Jersey, and currently Visiting Fellow at the Center of International Studies, Princeton University. Author of *Water and Power: the Politics of a Scarce Resource in the Jordan River Basin* and several related articles, her research focus has been on transboundary resource disputes in protracted conflict settings. She is on the Executive Committee of the Environmental Studies Section of the International Studies Association. She was a participant in the 'Environmental Change and Acute Conflict' joint project of the University of Toronto, Peace and Conflict Studies Program and the American Academy of Arts and Sciences (1990–92). She has also contributed to the ongoing 'Environmental Change and Security Project' of the Woodrow Wilson Center. Her current work is on economic crisis and political breakdown in oil-exporting states.

Elizabeth L. Malone is the manager of the Global Climate Change Group at the Pacific Northwest National Laboratory, Department of Energy, Government of the United States. She is co-editor (with Steve Rayner) of Human Choice and Climate Change, a multivolume assessment of social science research relevant to global climate change.

Richard A. Matthew is Assistant Professor of International and Environmental Politics in the School of Social Ecology at the University of California at Irvine and Director of the Global Environmental Change and Human Security Research Office at UCI. He has published numerous articles on environmental issues, ethics in international affairs and international organization, and is the co-editor of *Contested Grounds: Security and Conflict in the New Environmental Politics*. He has worked with the Foreign Services Training Center, NATO, USAID, CIDA, DOD, DIA, Department of State, WRI and other organizations on projects related to environment and security. He is on the Executive Committee of the Environmental Studies Section of the International Studies Association, the Advisory Boards of AVISO, the *National Security Studies Quarterly* and the *Journal of the Environment*, and the Science Committee of IHDP/GECHS.

John W. Mentz is a Treaty Manager in the Office of the Secretary of Defense, Government of the United States. He represented the Department of Defense at negotiations for the Open Skies Treaty, the Conventional Forces in Europe Treaty, and the Nuclear and Space Talks.

Steve Rayner is Chief Scientist with the Global Climate Change Research Group at the Pacific Northwest National Laboratory (PNNL) in Washington, DC. Before joining the PNNL in 1991, he was Deputy Director of the Global Environmental Studies Center at Oak Ridge National Laboratory. He has served with a number of institutions concerned with applied social research, risk management, and the human dimensions of global change, and published extensively in these, and other related, areas. He has advised the European Commission, as well as the national global change research programs of the United States, the United Kingdom, and the Netherlands. In addition to working at the PNNL, he also teaches in the Science, Technology, and Society Program at Virginia Tech.

Galina Sergen is a graduate fellow at the Pacific Northwest National Laboratory in Washington, DC.

Brian R. Shaw is the manager of the United States Department of Energy, Pacific Northwest National Laboratory's Center for Environmental Security. The Center is a part of the Arms Control and Nonproliferation program of the National Security Division, and was created to investigate the environmental components of regional tensions. A geologist by training, Brian Shaw's research and writing focus includes an array of regional security issues, among them nonproliferation and arms control, international impacts of nuclear fuel cycles, natural resource appraisal, energy security, and regional environmental management.

Michael Thompson is a social anthropologist. He is director of the Musgrave Institute, London, an adjunct professor in the Department of Comparative Politics at the University of Bergen, Norway, and a senior researcher in the Norwegian Research Centre in Organization and Management, Bergen. Author of many books and articles, his current interest is in the democratization of decision processes in areas (such as risk management, environment and development in the Himalaya, technology, and global climate change) that have tended to be treated as merely technical.

Aaron T. Wolf is an assistant professor of geography in the Department of Geosciences at Oregon State University. His research focus is on the interaction between water science and water policy, particularly as related to conflict prevention and resolution. He has acted as consultant to the US Department of State, the US Agency for International Development, and the World Bank on various aspects of transboundary water resources and dispute resolution. He is author of *Hydropolitics along the Jordan River: the Impact of Scarce Water Resources on the Arab–Israeli Conflict*, and a co-author of *Core and Periphery: a Comprehensive Approach to Middle Eastern Water*. Wolf co-ordinates the Transboundary Freshwater Dispute Database, an electronic compendium of case-studies of water conflicts and conflict resolution, international treaties, national compacts, and indigenous methods of water dispute resolution.

1
Introduction and Overview

Miriam R. Lowi and Brian R. Shaw

Environmental security has, over the last decade, emerged as a topic of discussion and vibrant debate. Largely in response to the end of the Cold War, the termination of superpower rivalry, and the explosion of nationalisms, the academic and policy communities have sought to revise their conceptions of security to include non-military threats and non-state actors, so as to more effectively explore and understand underlying issues that impact new (international) relationships. In keeping with this conceptual shift, the natural environment, and anthropogenic alterations of it, have been identified as important elements in the relations among states, communities, and individuals.

Indeed, the call for a redefinition of security preceded the end of the Cold War. As early as 1977, the environmentalist Lester Brown made the first, largely unheeded, plea. Then in 1983, Richard Ullman, in a seminal article, reiterated and refined that plea, calling for a broadened understanding of threats to security to include any action or sequence of events that 'threatens . . . to degrade the quality of life for the inhabitants of a state', or significantly constrains 'the range of policy choices available to a state or to private, non-governmental entities (persons, groups, corporations) within the state'. (133)

While the study of environmental politics that emerged as a research agenda in its own right some ten years ago has drawn inspiration from both Brown and Ullman, it is the work of Jessica Tuchman Matthews (1989) that led the way. In a passionate article written on the heels of the fall of the Soviet Union, she illustrates the ways in which environmental factors – be they of a planetary, national, or sub-national scale – threaten the security of states and of peoples. Hence, she argues, not only must we broaden our understanding of security to include previously overlooked variables, but we must also promote and implement policies that are 'environment-friendly', and protective of the planet, of humanity, and of national states.

What, however, do we mean by 'environmental security'? It is a contentious concept, often used in different ways by different disciplines and communities (Deudney, 1990, 1991; Goldstone, 1996; Homer-Dixon, 1996; Levy, 1995a; Porter, 1996). While this is often confusing, it illustrates, in fact, the sheer breadth – in terms of foci, discursive and ideological predispositions, disciplinary and other concerns – of the new environmental politics (Deudney and Matthew, 1999).

Several broad sets of questions have guided the research agendas of the 'environmental community'.[1] First, is the environment a security issue? How does the natural environment, or changes to it, affect security conditions? Moreover, if we accept that there is a connection between the natural environment and security, whose security are we most concerned about, and whose security is most at stake – the state's, people's, or the biosphere's? Who are the 'victims' and who are the 'perpetrators' of environmental change? And what is the relationship between them in historical, spatial, and socio-economic terms?

Second, given the traditional focus of security studies on conflict and war, is there a relationship between environmental factors – that is, conditions of, or changes to, the natural environment – and violence? If environmental change provokes conflict, does it do so at the inter-state or the intra-state level? Do states go to war because of environmental factors? Can civil strife be easily traced to environmental conditions? How can we determine whether a country has an interest in, and the means to, intervene effectively in areas where environmental factors are generating or exacerbating conflict and violence? Alternatively, could conditions of, or changes to, the environment encourage cooperation, rather than conflict, between states or communities?

Third, what are the most effective venues for addressing environmental security issues? Wherein lie the strengths and capabilities of the diverse array of traditional and emergent environmental 'activists'? Consider, for example, the United Nations system and its various specialized agencies and programs: UNDP, UNEP, UNCED; regional organizations such as NATO; non-governmental organizations such as Greenpeace and the World Wildlife Fund. What are the prerequisites for the formation of international environmental regimes?

And what has been the record of international negotiation over environmental concerns? Should we encourage the 'greening' of militaries? Are they credible as environmental activists? Under what circumstances would political and institutional reform be the most appropriate mechanism for mitigating environmental constraints? How can empowering populations and enlisting their participation assist in the formulation and implementation of effective response strategies? Does the mitigation of environmental constraints require that we address conditions of inequality, marginality and poverty?

Structure of the book

In this volume, authors address the above sets of questions. We offer perspectives; there are no definitive answers. We are an interdisciplinary community, composed of both academics and practitioners. We have different backgrounds and orientations. Our aim in writing this book is threefold: first, to move beyond the general discussion on the nature of security and the efforts to prove a link between environmental variables and violent conflict;[2] second, to highlight and explore the complexities of the relationship between environmental variables and security conditions; third, to continue the dialogue and re-focus the debate among disciplines, perspectives, and between academics and practitioners. We hope, in this way, to contribute to the accumulation of knowledge and problem-solving capacities.

The book is divided into two sections, with some degree of overlap between them. The first is concerned primarily with analytical frameworks for addressing the relationship between the environment and security. The second explores aspects of the relationship through case studies and within particular frameworks. There is no effort, in the architecture of the volume, to measure the contribution of the academy relative to that of government, or vice versa. To wit, several of the contributing practitioners are scholars in their own right, while most of the academics have consulted for government, and other, agencies.

The notion of national security has commonly been expressed through the role of the military. Such things as securing borders, projecting force and influence, and responding to national threats and adversaries are roles traditionally assigned to the armed forces. For these reasons alone, the ability to defend and protect national interests defines the persistent militarized understanding of security.

However, one of the results of the post-Cold War transformations has been, as mentioned above, a willingness to incorporate non-traditional security issues as central concerns of national strategies. In their chapter, Mary Margaret Evans, John W. Mentz, Robert W. Chandler and Stephanie L. Eubanks highlight the significance of this change for, and its effect on the evolution of thinking of, the defense establishment in the United States. They describe not only key policy changes, such as former Defense Secretary William J. Perry's preventive defense process, but also a deeper, and perhaps more refined understanding of the multifaceted and multidimensional nature of both national and regional security. To build conditions of peace, they propose a revamped policy process within the US government to coordinate security decisions among the various concerned agencies and departments. By re-tooling the interagency process to include experts on the multiplicity of factors that 'drive' (national or regional) insecurity, the broadened understanding of security can be institutionalized within the system. This should make it possible to identify potential problem areas more quickly and prepare responses more effectively.

Richard A. Matthew agrees that environmental factors can and should be integrated into traditional security affairs insofar as environmental change threatens national interests and, hence, becomes relevant to the conventional mandates of military, and other related, apparati. However, he cautions that doing so involves certain risks. For one, environmental change affects societies in a variety of ways; while in circumstances of limited capabilities, it may contribute to conflict, elsewhere it can stimulate technological innovation and/or inter-state cooperation. Traditional security institutions may not be equipped to identify the possibilities for and promote the latter. Moreover, a reliance exclusively upon security institutions to address environmental problems and their consequences limits severely the range of possibilities for, and efficacy of, response strategies. Because of their complex and often transnational character, environmental problems, he contends, are most effectively coped with through education, research, data sharing and multilateral cooperation. Matthew, therefore, appreciates the linkages between environmental change and conventional security affairs. He insists, though, that because of the diverse sources and impacts of, and means to address, environmental change, we need to move beyond a focus on the nation.

There is a group within the environmental community that identifies the individual and, by extension, the collectivity, as the referent object of security. In contrast to the statist conception and its attention to national sovereignty and the integrity of borders, at issue here is the welfare of the individual (and the collectivity): the protection from (threats of) abuse, dislocation, starvation, poverty, and illness, and the guarantee of civil liberties. Security in this context has as its primary goal to bring the resources of civil society to the benefit of the individual.

Steve Rayner and Elizabeth L. Malone suggest that if security is about the ability of the social system to resist the impact of a wide variety of disruptions, then security refers to the capacity for 'societal resilience'. Stated differently, the capacity to respond effectively to environmental security threats can be traced, to a significant degree, to institutional development at the level of society or community. That being the case, and building on research on the topic of social capital, the authors argue that security – no matter its referent object – requires investment in a variety of societal institutions. This is so, not because of institutions' direct contribution to economic efficiency or technological effectiveness, but as a way of ensuring that societies develop and retain a minimum level of institutional diversity for effective strategy switching as new challenges arise; institutional diversity provides societies with a plurality of viewpoints and mechanisms to draw upon. Hence, the most effective means of addressing security concerns is, initially, to promote the emergence of a strong civil society.

Steve Lonergan notes that human security focusses on individual or collective human perceptions and evaluations of actual and/or anticipated environmental conditions as sources of insecurity. He maintains that perceptions of the natural environment, and the way we exploit the environment, are socially, economically and politically constructed. As such, the human security perspective – of which he is a proponent – is a participatory one, where 'those being researched become the researchers, and vice versa.' Lonergan aims to clarify the relationships between environment and security, human security and sustainable development, and the overlaps between the two sets. He argues that environmental problems must be addressed from a perspective that encompasses both world poverty and issues of equity. This is because poverty and inequity are two of the key factors contributing to tension and insecurity throughout the world. Finally, he insists that 'space' is a critical variable of environment and security relationships. Moreover, as spatial dynamics shift both globally and locally, they become even more central to these relationships and discussions about them. Indeed, the significance of space – the location of environmental security challenges – is especially noteworthy, not only for our comprehension of the complexities of these challenges, but also, in the formulation of response strategies: Lonergan maintains that environmental and security concerns are dealt with most effectively at the level where the knowledge base and the effects are greatest.

The significance of the spatial dimension in so-called environmental security relationships is demonstrated further by Simon Dalby. A dissenting voice in the environmental community, Dalby criticizes the disquieting absence, or misrepresentation, of the spatial dimension in discussions of environmental security. He points out that more often than not, it is the advanced industrialized countries – the North – that, in their quest for resources, have provoked environmental changes in lesser developed countries – the South, with debilitating conseqences for the natural environment and human security. Not only is there a profound spatial – and socio-economic – dimension to environmental change and (in)security, with the North (rich) and the South (poor) sparring with each other, but also, the relationship between them represents a centuries-old pattern: a pattern that has been accompanied by conflict and violence. Dalby, therefore, takes exception with what he views as the elitist, ahistorical, and de-contextualized character of most discussions about environmental security. He makes a plea for a fundamentally revised research agenda: one that is sensitive to the complexities of relationships, and incorporates the critical elements of place, time, and *rapports de force* into the analysis of environmental change and the politics of security.

The second section of the volume, that presents case study material within particular analytical frameworks, begins with Robert E. Bedeski's application of non-traditional security concerns to the continued survival of South Korea. In his chapter, Bedeski points out that until recently, Korea evaluated its security solely in conventional terms: with a focus on the integrity of its borders, the recognition of its geopolitical sovereignty, and the extent of its coercive capabilities relative to its neighbors. However, as in Chapter 2, where Mary Margaret Evans and her co-authors outline the appropriateness, in the post-Cold War environment, of incorporating non-traditional security issues in national strategies, Bedeski shows that South Korea, in response to rapid industrialization and political development, has begun to modify its conception of security to include non-traditional challenges: among them, the state of the economy, environmental degradation, resource availability, drug trafficking and human rights abuses.

The relationship between natural resource availability and the security of states and peoples is dealt with explicitly in the next two chapters and with a focus on water. Aaron T. Wolf and Jesse H. Hamner begin the discussion by providing evidence to debunk the 'water wars' hypothesis. They demonstrate unequivocally that the historical record is one in which inter-state cooperation over the distribution and utilization of (scarce) water resources is far more common than is inter-state war. Conflicts over water do arise and are numerous, but more often than not, they are resolved through negotiation and ultimately, with the signing of treaties. The authors outline, as well, a variety of mechanisms that can be brought to bear on the disputants to facilitate dispute reduction and water sharing.

Miriam R. Lowi also addresses the 'water wars' hypothesis, this time through evidence from Middle Eastern and South Asian river basins, and in response to the preoccupation of the 'second wave' of environmental security research. In fact, her contribution is structured as a series of responses to a set of questions that has dominated much of the debate to date: first, should environmental change be considered a national security concern? Second, is there a positive relationship between environmental change and acute conflict? Third, are the effects of environmental change in different parts of the world so critical that, in the absence of immediate action, irreversible humanitarian disaster is inevitable? Her responses reiterate the need for studies of the relationship between environment and security that are historically informed, context-specific, and sensitive to the complexities of context. Moreover, she makes a plea for a less narrowly-focused research agenda, and far greater attention to the importance of political and institutional reform in the attenuation of environmental constraints.

The remaining two chapters develop further the theme of institutional reform and its role in environmental management. In discussing the case of the dessication of the Aral Sea, Galina Sergen and Elisabeth L. Malone note that because environmental problems derive from human institutional arrangements, the stability of local and regional institutions is critical for resolving environmental issues and maintaining regional stability. Moreover, to address concerns regarding enviornmental security, and specifically, riparian issues in the Aral Sea basin, Sergen and Malone propose 'a risk-oriented framing that emphasizes pluralism as an essential element'; risk analysis, as they describe it, can help solve environmental conflicts. Furthermore, like Wolf and Lowi, they reject the identification of riparian issues solely in terms of their potential for conflict. That, they insist, ignores the whole range of ways in which states and peoples resolve the problems they face.

The final chapter of the second section, and of the volume, is similar to the final chapter of the first section in that it too turns the environmental security debate on its head. While Simon Dalby contends that in our efforts, we have ignored, dismissed or misconstrued place, time, and the relative capacities of concerned actors, Michael Thompson argues that the environmental security community in general – and more surprisingly, the ecosystem and human security communities, in particular – have ignored, dismissed, or misunderstood the very objects of their avowed concerns: human populations. Drawing upon the experience of environmental degradation in the Himalaya, and the efforts of international organizations to address various aspects of this issue, Thompson notes that it is the institutional capacities of the people of the Himalayan region, and the appropriateness of the state's interactions with these capacities, that define the efficacy of interventions to resolve environmental constraints. He insists that brute interventions on the part of paternalistic states and ill-informed organizations are themselves major threats to the security of those they profess to be helping. In essence, he, like several others in this volume, is calling for a more participatory framework for addressing environmental challenges: a framework that includes the building of institutional capacity and the empowering of local populations in the formulation and implementation of strategies to improve their own lives.

Notes

1 In addition to environmental security, two other broad issues guide environmental research, policy-making, and activism; they are environmental ethics and sustainable development. Although linked to environmental security, this volume devotes somewhat less attention to them.
2 Marc Levy (1995a, 1995b) labeled the general discussions on the nature of security and the role of environmental degradation as a contributor to insecurity and conflict as the 'first wave', and the subsequent efforts to demonstrate a link between environmental variables and conflict as the 'second wave'.

Part I
Discourses

2
The Changing Definition of National Security

Mary Margaret Evans, John W. Mentz, Robert W. Chandler and Stephanie L. Eubanks

We are living in an age of revolution on many fronts – in politics with the end of the Cold War, in economics with the growth of global trade, and in technology with the leap-ahead explosion of information systems. This changing environment provides a rich array of possibilities for addressing national security in its broadest terms. As explained by former Secretary of Defense William J. Perry, '... the United States... has a unique historical opportunity, the opportunity to prevent the conditions for conflict and help to create the conditions for peace.' In a May 13, 1996 address at Harvard University, Secretary Perry called this 'preventive defense' and offered a definition: 'Preventive defense may be thought of as analogous to preventive medicine. Preventive medicine creates the conditions that support health, making disease less likely and surgery unnecessary. Preventive defense creates the conditions which support peace, making war less likely and deterrence unnecessary' (Perry, 1996, 1).

The concept of preventive defense was first introduced by Secretary of State George Marshall more than 50 years ago. In embracing the idea, Secretary Perry completed a full conceptual circle back to Marshall's broad vision of security. Perry's observations bridge the past and present conceptions of security and provide a starting point for analyses of future efforts to promote peace and stability. He has challenged us to step forward with an open mind to define the prerequisites for tomorrow's security in regions around the world.

Prior to World War II, the dominant view of security in the United States centered on insulating the Western Hemisphere from outside influences. This perspective was reinforced by the tradition of American isolationism, disillusionment with the result of World War I (the 'war to end all wars'), a belief in economic self-sufficiency, preoccupation with attempts to recover from the great depression, and deep-seated concerns about becoming involved in wars in Asia and Europe where America had no visible stake (Gaddis, 1987, 22).

This American mind-set began to change with the Japanese attack on Pearl Harbor and the subsequent American entry into the Second World War. Walter Lippman, in one of the most influential books written during the war, *Foreign Policy: Shield of the Republic* (1943), observed that America had too easily come to believe 'that a concern with the foundations of national security, with arms, with strategy, and with diplomacy, was beneath the dignity of idealists' (as quoted in Gaddis, 1987, 23). The issue facing America in the post-war era was often expressed in stark terms: 'Shall we protect our interests by defense on this side of the water', Princeton University professor Nicholas Spykman asked in 1942, 'or by active participation in the lands across the ocean?' (Spykman, 1942, 7).

A new approach emerged from fighting the war and preparing for the post-1945 peace – a doctrine of national security to explain America's new role in the world. American leaders were already using this approach by the end of the war to describe their version of the new peace. The idea of 'security against foreign danger', rooted in Madison's *The Federalist* during the late eighteenth century, testifies that these were not new thoughts in American political discourse.

After the end of World War II, the new concept of national security was a statement of America's leadership in the world. It connected US foreign policy and military affairs, and it postulated a dynamic interrelationship among the many political, economic, military, and other key non-military factors having an impact on US core interests abroad. Ultimately, it provided the rationale for the military containment of the Soviet Union and questions about how stable different distributions of power were likely to be (Waltz, 1969, 312–13).[1] Initially, the political, economic, demographic, and other principal factors associated with the well-being of the people and the viability of their democratic governments loomed as major concerns. When US policymakers looked at Western Europe between 1945–47, they saw before them a deep economic crisis with potentially drastic political ramifications. Washington used American economic power to help revive Western Europe's economies and create an environment hospitable to democracy and capitalism. The creation of NATO provided the military stability that gave the economic measures time to work, while containing the Soviet Union and the spread of communism (Yergin, 1977, 194, 303, 327). National security was recognized as an all-embracing doctrine to help build the conditions for peace.

General George C. Marshall, the US Secretary of State, was convinced that the Soviet Union was deliberately trying to retard Europe's recovery to further its own political goals. Such an assault on building the future security and economic well being of European countries could not pass unanswered. In a brief address at Harvard University on June 5, 1947, Secretary Marshall outlined the US response: 'The truth of the matter is that Europe's requirements for the next three or four years of foreign food and other products – principally from America – are so much greater than her present ability to pay that she must have substantial help, or face *economic, social and political deterioration* of a very grave character', Secretary Marshall warned. 'The remedy', he said, 'seems to lie in breaking the vicious cycle and restoring confidence of the people of Europe and the economic future of their own countries and of Europe as a whole. The manufacturer and farmer throughout wide areas must be able and willing to exchange their products for currencies the continuing value of which is not open to question' [emphasis added] (Weiss, 1995, appendix A).

Secretary Marshall's first use of the new national security doctrine was political and economic, not military. Post-war domestic chaos in Europe threatened to unleash forces that could undo the successes of the wartime effort. He knew that a lasting European stability had to rest on firm political, economic, and social foundations. Marshall also knew that lasting stability must grow from within Europe, not as a response to outside challenges. Without free political institutions, vibrant economies, and new military alliances, the cycle might progress toward another major conflict – or sublimation into the Soviet empire. The Marshall Plan sought to defend US interests in Europe by addressing the underlying causes of instability. Secretary Marshall's recognition of the importance of non-military factors in the national security equation led to about $13 billion in US assistance to Western Europe between 1948 and 1951. This infusion of capital, coordinated and targeted at the areas where it could be most effective, rejuvenated the struggling economies and allowed the Europeans to begin to help themselves toward recovery.

Despite the success of the Marshall Plan, the inclusive concept of national security that flourished in the immediate post-war years was soon smothered by the series of military exigencies that defined the beginning of the Cold War. Several factors predicted that the superpower standoff would be the single main determining factor in Cold War national security policy and the basis for global stability: the formation of NATO and the Warsaw Pact; the US recognition that the sizable Red Army forces positioned in Eastern Europe were there for the long haul; the communist takeover of China; Soviet acquisition of nuclear weapons; Soviet challenges to Western occupation rights in Berlin; and the Korean War.

Under the increasing pressure of the Cold War, the enlightened concept of American national security became more narrowly focused on anti-communism, containment, confrontation and intervention. Over the following decades, the non-military elements of security – involving political, economic, cultural, demographic and environmental factors – took a decidedly secondary role compared to the main ideological and military themes dominating the Cold War. In the great 'zero-sum game', the two superpowers extended their influence around the world – each in a quest to recruit more allies than the other. There could be no purely regional conflicts. Globally, security and stability were reduced to a matter of quantifiable military parity between the United States and the Soviet Union. Regionally, the same equations defined security between their respective allies and surrogates. It is a telling sign that the Cold War ended in 1991 largely due to the economic breakdown of the Soviet Union, not an ultimate military conflict. The Soviet bloc's support infrastructure simply could not sustain its military commitments.

The crumbling of the Berlin Wall and dissolution of the Soviet Union changed the international security environment dramatically. Former members of the Warsaw Pact and Soviet Union became free and independent of totalitarian control for the first time since the end of World War II. After 45 years of communist rule and centralized planning, the economies in Central Europe and the newly independent states (former Soviet republics) were in shambles. For two generations, private institutions, banks, insurance companies, and shipping firms had not existed, and commercial ties had disappeared. Free market institutions and practices had to be introduced from scratch. Factories and machinery were outdated. Finance capital was in short supply and assistance was needed in the agricultural sector and rural areas for such improvements as land reclamation, drainage, replanting cleared land and livestock improvement. The

Soviet emphasis on heavy manufacturing had ignored environmental problems that now had to be addressed. Public health, schools, and telecommunications needed restoration. In some countries, ethnic strife and millions of displaced persons added to conditions for instability. Thus, the Central European countries and the newly independent states have been faced with the enormously complex tasks of redefining their political institutions, restructuring their economic systems, and recreating national identities to include ethnic and religious minorities (*Economist*, 1995, 5). These states have been preoccupied with rebuilding and reorganizing within their boundaries, and have not generally been inclined or able to make sizable military investments. Thus, military parity has lost its predominant role as the major determining factor in a region's stability.

Elsewhere in the world, the decrease in superpower influence and control since the Cold War, and with it, the lessened threat of world war, have allowed new or formerly suppressed forces to endanger security on a regional scale. The dramatic revival of religious and ethnic rivalries; a population explosion in poorer states accompanied by increased migration to richer ones; powerful transnational forces of terrorism, drug trade, and organized crime; and rising pollution all present challenges that do not fit into the mold of Cold War security policymaking. (Ellsworth, 1997, 1–4). These threats to security manifest themselves at the regional and sub-national level. Regional conflicts present a host of additional complications for policymakers. They involve issues of sovereignty and statehood, and are often fought with guerrilla or terrorist tactics rather than by organized troops. Outbreaks of military violence are often accompanied or precipitated by a humanitarian crisis that needs to be resolved simultaneously. Negotiations become much more complex when the parties are no longer just state governments, but also ethnic groups, tribes, clans, separatist factions, and religious leaders – none of whom may have control over the fighting groups. In many cases, regional conflicts may prove more difficult to moderate than anything we dealt with in the Cold War, because they are based on disputes that have festered for centuries.

Our ability, as a nation and as an international community, to address these regional challenges to security will in large part determine whether we are able to reap the benefits of the demise of the Cold War and enjoy peace and stability in the twenty first century. To do so successfully, we will have to revive the post-World War II meaning of national security to embrace Marshall's concept and address all of the military and non-military factors at work in the post-Cold War era. In order to practice preventive defense, we must recognize the non-military factors in a conflict, create a process for assessing possible courses of action, and apply that process to mitigating the causes for conflict in troubled regions around the globe where it is in the interests of the US to do so.

In order to realize Secretary Perry's vision of security, we need to find ways to mitigate the causes of conflict in troubled regions around the globe. A three-step analysis seems to be a useful starting point for thinking about building the conditions of peace and setting the initial foundations for a broader, all inclusive analytical process.

- Assessing the Factors Driving National Security. Assess the impact of six key factors that remain unresolved, and in some cases unaddressed, which could lead to instability and war: military, political, economic, culture, demographic, and environment.
- Creating a Preventive Defense Process. Assess the Cold War negotiating process that dealt with the *symptoms* of conflict (the military arsenals of the United States and the Soviet Union), so that we can develop useful measures for a unified US government preventive defense process that will address the root *causes* of instability.
- Practicing Preventive Defense. Use the preventive defense process to identify potential regional security problems and address them in a coordinated manner, strengthening peace and pre-empting armed conflict.

National security factors

The states of Central and Eastern Europe have become the leading example – a test case – of the need to broaden our definition of security. However, trends are evident in all parts of the world that lead to the same conclusion. Political, economic, cultural, demographic and environmental factors are not incidental to the definition of security; they are an integral part of it. Understanding how these key forces affect security, and how to address them to improve security, is the challenge in the concept of preventive defense. The case examples for each of the six interrelated and mutually reinforcing variables that follow have been selected to illustrate the challenges involved in building the conditions of peace.

The military

A perceived military balance on the Korean peninsula from 1953 through the end of the Cold War maintained stability in an atmosphere of intense hostility and has extended into the new era. North Korea's animosity toward the Republic of Korea has continued through several military challenge and response cycles since the 1950–53 Korean War. For decades, South Korea, with the US as its ally, was ensured of a rough military parity with the North to keep Pyongyang at bay. Each side developed infantry forces supported by tanks and artillery. Later, North Korea developed a significant special forces capability. In response, South Korea and the United States used other means, including 'defense in-depth'[2] and use of air forces and other means of indirect fire, to neutralize the potential effects of the North's commando operations. In the 1980s, the North developed a large force of armor and artillery to provide greater mobility. The South Koreans and Americans turned to advanced technologies to neutralize the North's advances in mobility, including enhancements in intelligence collection and analysis to enable precision targeting. New operational concepts accompanied this shift in capabilities, placing the North's armored forces and artillery at severe risk. Pyongyang countered with the development of biological and chemical agents for attacks against airfields to relieve the pressure on its tanks and artillery (Bennett, 1996, 2–9).

Today, North Korea possesses about 5000 tons of chemical weapons, including mustard and blister agents – or enough to bring widespread destruction to South Korea (Kim Sang-Beom, 1997). Pyongyang is suspected of having anthrax and other biological agents as well. When combined with the North's prodigious Nodong and Taepodong missile programs, which produce about 100 missiles per year, the North Korean threat covers the entire territory of South Korea as well as extensive areas of Japan. According to military defectors from the North, the strategy is to inflict more than 20 000 American casualties in the region, including those in Japan, Okinawa, and Guam (Gertz, 1997).

Meanwhile, the country's failing food distribution system has resulted in tens of thousands of North Koreans dying from starvation and starvation-related illnesses, most occurring in the northern part of the country in the fall of 1996. North Korea has requested and received emergency relief from the international community. At the same time, the country's gross domestic product dropped 5 per cent annually from 1992 through mid-1997. Yet, Pyongyang has continued to invest its scarce resources in developing and maintaining its armed forces, including its chemical and biological warfare and missile programs (Snyder, 1997, 2; US, Office of the Secretary of Defense, 1997, 4).

Will the 1998 Korean four-party talks – North Korea, South Korea, China and the United States – provide a pathway toward a formal end of the Korean War and then reunification? Will the North change its attitude toward South Korea? Should the South and the other members of the talks wait for the North to change? Will military parity preserve stability or will an increasingly desperate North Korea lash out before it is cemented into a position of military inferiority? How could the other stability factors in the North–South standoff facilitate a negotiating process leading to a peaceful resolution of differences?

Politics

The Central European and newly independent states have been preoccupied with establishing their political identities. Tensions between the old guard communists and reform-minded democrats have produced an uneasy standoff between the main political contenders in several countries. In some countries, bureaucrats from the former communist regimes have burrowed into the warrens of government and agitate for the return of the 'good old days'. In others, they have been reborn as 'socialists' and still sit close to the centers of power. For the democratically-minded, establishing linkages with the West through the United Nations, European Union, and NATO have been ways to distance themselves from their communist past and to open more avenues of Western aid.

Armenia is an example of a country where the political transition has not been easy. This country of about 3.7 million people is slightly larger than the state of Maryland. Almost from the moment it gained independence, the country faltered and stumbled toward autocracy. A victory in a six-year war with neighboring Azerbaijan over the Nagorno–Karabakh territory was good for national pride, but it crippled the economy. Presidential elections in 1996 were suspect, prompting a 'velvet coup' and the scheduling of new elections (Williams, 1998, A16).

The Armenians have options open as a political realignment is underway in the Middle East and the Caucasus. A Greek–Syrian–Russian bloc has added Armenia to its developing alliance. Turkey and Israel oppose the arrangement. Athens has also invited Azerbaijan to join the bloc, but Baku is wary of Greece's collaboration with Armenia in the military sphere. Oil pipelines from Central Asia are intrinsically linked to the ongoing political realignment. The natural pipeline route from Azerbaijan would run through Armenia, but it has been routed around the country due to Armenia's differences with Azerbaijan and a closed border with Turkey (Abruzere, 1998, 1).

How could the international community try to open a dialogue between Armenia and Azerbaijan to find a working arrangement despite their mutual mistrust? What kind of a negotiating process might be best used in addressing the equities of all of the concerned parties? How could the other factors of stability be used to facilitate this process?

The economy

Economic disaster confronted several countries in the early 1990s with the loss of financial support from the Soviet Union and the shock of market reforms. The lack of modern communications and transportation infrastructures and a simple lack of know-how plagued their early economic rebuilding efforts. Hyperinflation, market instability, and a lack of diversified industrial production all hindered attempts to introduce market economies. Multilateral and bilateral assistance from Western institutions and countries through the 1990s helped to reorient some of these economies along free market lines, although holdover supporters of central planning continue to slow the efforts of the reform minded. The delays and difficulties of the transition have left some countries in a state of flux with a dysfunctional hybrid economic system.

Central Europe provides several examples of the influence of both economic success and failure on security and stability. The Czech Republic, Poland, Slovenia and Hungary moved quickly toward privatization and rapid economic growth rates.

Albania, on the other hand, collapsed into anarchy in the spring of 1997 as a direct result of economic mistakes. Albania entered the post-Cold War period as one of the poorest countries in Europe. For five years, the first noncommunist government elected since World War II failed to introduce functioning democratic and legal institutions, and there was a lengthy struggle between political parties. The economy, however, was one of the fastest growing in Central Europe. Per capita income more than tripled from 1992 to 1995, unemployment fell from 40 per cent in 1992 to less than 13 per cent in late 1996, and inflation dropped drastically.

The impressive rates of change, however, masked the difficulties of the lack of a proper private banking system and a dysfunctional stock exchange. Rather than put their money into a shaky banking system, savers invested instead in pyramid schemes offering extraordinary high rates of interest. New deposits were used to pay interest rates of previous investors while the companies running the pyramid schemes were cited as examples of Albania's economic success. The results were inevitable. When the pyramid investment schemes collapsed in late 1997, the country descended into anarchy. Arms depots were looted, public order collapsed, and banks were robbed. Public buildings were destroyed. About 1500 people died, mostly from the haphazard firing of looted weapons. Criminal gangs quickly took advantage of the breakdown of order. Albania seemed poised on the edge of civil war (Schmidt, 1998, 127–31).

A multi-national task force arrived in Albania in April 1997 to guarantee the delivery of humanitarian aid and stabilized the situation. Meanwhile, the political climate is not conducive to fair and free elections, and the Albanians do not have the resources necessary for economic revitalization. (Xhudo, 1997, 260–5).

The problems of transition from centralized planning to democracy and a market economy in Albania were clearly visible to the international community. Could something have been done by the European and American assistance programs, as well as international financial institutions, to head-off the growing crisis? How could the international community have prevented the collapse of the banking system and stock exchange, thus, precluding the foolish pyramid schemes, without interfering in the internal decision-making of the country? Could not the funds now being spent to support the multinational task force in Albania have been put to use better in restructuring the economy before the crisis occurred? How could other stability factors be used to mediate the situation?

Culture

Cultural identity, 'including crosses, crescents, and even head coverings', as well as other symbols of culture, count in the post-Cold War world, according to Harvard University's Samuel P. Huntington. 'For peoples seeking identity and reinventing ethnicity', Huntington warns, 'enemies are essential... culture and cultural identities... are shaping the patterns of cohesion, disintegration, and conflict in the post-Cold War world' (Huntington, 1996, 20). While Huntington may overstate the case, the collapse of communism has allowed the resurgence of ethnic and religious identities once suppressed. Ethnic and religious differences in Bosnia, elsewhere in Central Europe, and in the newly independent states, for instance, have undermined the cohesion necessary for making needed political and economic reforms. In some of these areas and in Africa, these disputes have erupted in violent conflict. Cultural disagreements, particularly those of a religious nature, also are an underlying factor in some of the world's most longstanding and dangerous conflicts, obvious examples being India and Pakistan, and Israel and its Arab neighbors.

Macedonia is all too familiar with the ways in which cultural differences can disrupt a state. To many nationalists in the Balkans, Macedonia should be torn asunder. Bulgaria and Serbia both lay territorial claims to the small state, while Greece decries the use of the name Macedonia to describe the country and its people. Macedonia is a multi-ethnic state of some two million people. Ethnic Macedonians account for about two-thirds of the population. Albanians are the next largest group (23 per cent), followed by Turks, Roma (also known by the derogatory term 'Gypsy'), Serbs and other small groups such as Vlachs, Torbeshi (Macedonian Muslims), Croats, Bosnians and Bulgarians. Macedonia is no melting pot; the various peoples live in closed, separate communities. Members of the diverse ethnic groups take pride in belonging to their particular community and sharing their language and customs. They are dedicated to preserving their individual identities.

Macedonia's track record in dealing with minorities is uneven. Conflict has occurred with the government's treatment of Vlachs, a nationalistic Serbian minority backed by Belgrade. The Macedonians, who are ethnic Slavs and Orthodox Christians, find the Albanian minority, non-Slavic and mostly Muslim, to be the greatest threat to security. The Albanians have traditionally lived in largely homogenous ethnic clusters. They feel alienated from Macedonian society. Macedonians and other ethnic groups feel unwelcome in the areas where Albanians are concentrated. The 1998 uprising in Kosovo by Albanians may fuel Macedonia's fear of a 'Greater Albania' that would include Albania, Kosovo and western Macedonia.

Albanian politicians argue that they simply want the same opportunities as those enjoyed by the ethnic Macedonians. The Albanians want to govern themselves within the sovereign context of Macedonia, be allowed to display their cultural flag (not the Albanian state flag), use their native language in state commerce, use Albanian as the language of instruction in their schools, and live free from fear of police intimidation and violence. Thus, in Macedonia, the majority group feels threatened by the Albanians' political and cultural desires (Perry, 1998, 119–26; Spolar, 1998, A17).

Will inter-ethnic fighting break out in Macedonia? Since disintegration would worsen the fragmentation already present in the former Yugoslavia, should the international Contact Group[3] encourage internal political resolution of ethnic minority issues? How might the ethnic Macedonians reach an accommodation with the Albanian minority? How can the political and other relevant factors be incorporated to help find the answer?[4]

Demography

Population growth in the developing world can affect US security interests indirectly in many ways, most of which are related to the distribution of resources, both within the affected area and between developed and developing countries on a global scale. The strain on an area's resources to support rapid growth may lead to high unemployment, environmental degradation, food shortages, or disease epidemics, all of which can lead to political instability and migration pressures.

In Central Europe and the newly independent states, the primary demographic challenge has been and continues to be the movement of populations from one state to another. Demographic forces following the breakup of the Soviet Union triggered the largest population upheaval since World War II. More than nine million people were displaced throughout Europe and Central Asia. The largest percentage of these were people who had been living outside their home republics in the Soviet Union. Facing an uncertain future and rising nationalism, particularly of an anti-Slavic bent in the Central Asian and Baltic republics, many moved back to their home states. Such mass migrations place a strain on the social and physical infrastructure and resources of the states involved (Goshko, 1996, A20).

The war in Bosnia produced one million refugees and nearly as many displaced persons who were either forced from their homes or fled to avoid the internecine violence. One promise of the Dayton Agreement[5] was the return of these refugees to their homes. Yet the parties have been slow to comply due to political, economic, social, legal, and security and personal safety obstacles.

At Dayton, the Contact Group considered the refugee return to be a matter of urgent concern. Annex 7 of the Dayton Agreement fully details the rights of refugees. Chapter 1 provides the script for return, identifying the key actions and directing various activities of the parties to the Agreement. Chapter 2 presents the back-up plan that outlines what should be done for refugees who cannot return to their homes and areas of origin. For those who had lost their homes or were too traumatized to return, Dayton established the Property Commission to ensure that victims were compensated for their loss. It was assumed at Dayton that most refugees would be of the Chapter 1 returns. The opposite has proven true. The Property Commission has become the primary mechanism for resettling refugees and displaced persons. Meanwhile, returnees from Europe would end up massed in just a few areas, creating the potential for social tension. High level political figures in Serbia, Croatia and Bosnia continue to promote ethnic separation. Displaced persons remain a problematic social factor and have often been used as the pretext for acts of intimidation and harassment against minorities.

The resettlement of refugees is linked intrinsically to the final outcome of the conflict in terms of territorial integrity, national identity, and sovereignty, and the disputants have been accused of manipulating the relocation of displaced persons to advance their political interests. The United Nations High Commissioner for Refugees (UNHCR) estimated, for instance, that 870 000 persons would return to their homes in 1996, but only 250 000 people actually resettled, 240 000 of which were in areas where their ethnic group was in the majority. Resettling refugees into minority areas is problematic from their personal security point of view. Settling these people instead in majority areas exposes the resettlement process to the vagaries of political interests of contending parties (Akan, 1997, 1–11 and US Department of State, 1998).

Bosnia holds important lessons for managing refugee returns in the post-Cold War era. What further steps might be taken to encourage the return of refugees? What process will best encourage an open dialogue among the concerned parties on resettlement of refugees and displaced persons? How can the stability factors – political, cultural, military, economic, environmental and demographic – be used to facilitate the process?

The environment

Not all environmental issues are significant in terms of national security; to claim so would present an unbounded agenda of environmental concerns and would obscure the very real environmental threats that do affect security. The storage and disposal of toxic waste products, particularly ones related to the manufacture of weapons, can be of paramount concern to a state's or region's stability. If the US has strong national interests in maintaining that region's stability, it would be considered a US national security issue. Worldwide issues such as global warming, ozone depletion, over-fishing or threats to biodiversity, clearly make the environmental agenda, but not all of these should be on the national security agenda. Inclusion would depend on the region and on US interests. In some regions, problems such as trans-border air pollution, water resource disputes, and shipment of toxic wastes between countries are closely tied to security. (Keller, 1996, 1–4).

How does one determine what environmental issues are security issues for the US? The after-effects of Soviet industrial and military practices in Central Europe are high among the examples of environmental problems with serious national security implications: disposal of radioactive materials at sea, dumping chemical munitions into the Baltic Sea, and the Chernobyl nuclear power plant disaster. In Haiti and sub-Saharan Africa, environmental problems plague economic restructuring programs. Deforestation in Haiti, for example, has led to an inability to preserve the national agricultural output at a subsistence level. The resulting deteriorating economic conditions lead to civil unrest and ultimately to illegal immigration to the United States.

Creating a preventive defense process

The questions posed for each of the six national security factors discussed above reveal some common elements or potential building blocks that might be useful for creating a preventive defense process: (1) each element presents a difficult problem on its own; (2) in each instance there is considerable overlap or common ground shared with one or more other factors; (3) each national security factor contains an inherent element for instability, sometimes engulfing other security factors; (4) each of the driving factors are open to human control and decision, opening the possibility of negotiations to resolve difficult issues; (5) in some cases, resources may be needed to 'prime the pump' of peaceful settlement of disputes; (6) the common denominator for each national security factor is political – the concerted action by government, drawing on past negotiation experiences, is needed; and (7) multi-national entities or key countries with an interest in a particular problem (eg the five-member Contact Group for Bosnia) may provide a useful framework for discussion between disputants.

Perhaps our experience with the arms control process during the Cold War offers some insight into how we can adapt national security policy to address these non-military factors. After numerous fits and starts in the 1950s, arms control developed over time as a useful and essential tool of Cold War policymaking. Since arms control could do nothing about the central ideological clash of the superpowers, it addressed the symptoms of the struggle: military escalation, the pursuit of a nuclear first strike capability, and militarization of space. One of the key lessons the US learned from the Cold War dialogue is that, with appropriate national will, the negotiation process works. The agreements successfully reached during this period include the Anti-Ballistic Missile Treaty (ABM-1972), Nuclear Non-Proliferation Treaty (NPT-1968), Intermediate-range Nuclear Forces Treaty (INF-1987), Conventional Armed Forces in Europe Treaty (CFE-1990) and the Strategic Arms Reduction Treaty (START-1991).

In its basic definition, arms control is about negotiation that leads to limits on the weapons of the other side and on our own military forces. Russia's SS-18 multiple nuclear warhead missiles, for example, are the most destructive weapons ever aimed at America. Under the START treaties (existing and future), all of these missiles will be eliminated and their destruction verified. This elimination of one of the greatest threats to the United States was made possible through negotiations and without a shot being fired.

We need to build upon the successful elements of Cold War negotiations, while adapting our process to address new security realities. The Cold War arms control talks succeeded in part because, in the US, an interagency process was established to promote sound decision-making. The relevant players within the US executive branch were identified and brought to the table, where they were able to describe their equities in a particular situation. The players researched the issues and discussed the pros and cons in an attempt to reach a consensus. Even if a consensus could not be reached, a final, informed decision was taken with an understanding of the equities involved. These tactics can translate directly into use for meeting new security challenges as well. The difference is the focus.

The Cold War is over, but we are still drawn to the familiar ground of dealing with the *symptoms* rather than the *causes* of conflict.

Can we shift our focus from symptoms to causes? An entirely new process is needed to address the causes of instability in the post-Cold War era. It will require a major shift in our basic security assumptions to reorient efforts to address the causes of insecurity broadly defined – the political, economic, cultural, demographic, environmental *and* military factors.

The problem with traditional arms control negotiations is that they only address situations where instability is manifested through a weapons buildup. Regions with political, economic or other non-military problems are ignored by the US security policy community until they become military and/or humanitarian crises. Even when these problems attract attention, the US government usually deals with them in a compartmentalized fashion. So, for example, environmental experts work on environmental issues, economists look at economic trends, and never the twain shall meet. If the US government recognized the security impacts of these factors, they could be analysed in a true non-compartmentalized interagency process. This process could go a long way toward facilitating preventive defense by bringing together experts in the various security domains and prompting them to craft solutions from a joint perspective. Such joint intervention, introduced at an early stage of the problem, might mitigate the issues so that troops are not required.

The task at hand is to create a preventive defense process whose purpose is to prevent war by creating the conditions for peace. All six of the main national security factors should be encompassed in a single process – not six independent frameworks where each of the national security factors are assessed individually. The process should address national security problems as a whole, relating all factors to each other. This does not mean that each factor should not be examined closely and in detail, but the focus should be on the interaction between the six main national security factors. The factors are intertwined and influence each other in complex ways, which makes considering them in isolation misleading and nonproductive. The degree of influence exerted by these key elements will vary in relation to each other from issue to issue and region to region, and they will differ in their relative contribution to solutions in particular problem areas.

The creation of a functioning preventive defense process based on the existing interagency process should sit high on the American national agenda. The first step needed is to identify the right departments, agencies, and offices of the US government that can contribute to a broadly based preventive defense approach. Convincing non-defense related organizations that they *do* have an equity in security affairs, and vice-versa, will be difficult. However, once this is achieved, the US government's ability to detect potential problem spots and get the right people looking at the issues should increase dramatically. An example of this approach is the recently formed interagency working group on counterterrorism. In an attempt to pull in all of the relevant interagency players, this group includes representatives ranging from the Federal Bureau of Investigation, to the Office of Management and Budget, to the Department of Health and Human Services. This sort of broad-based approach to interagency coordination should be a cornerstone of a workable preventive defense process.

After an issue has been identified, it must be worked through the expanded interagency process. One way to proceed would be to have US government experts on the relevant security factors prepare a series of white papers on how to address the issue from their particular perspectives. These papers and/or interagency discussions would likely provide information that simply would have been overlooked in a narrower approach. The preventive defense process would cut across government agencies and create an approach that integrates their efforts and provides a forum for exchanges of views. Once the facts are known pertaining to all of the relevant security factors, the players could debate pros and cons and make an informed decision on a course of action.

After the US government has formed a coordinated approach to an issue, it should try, as it does now, to garner support from other countries and organizations before taking action. Allies and partner countries are needed for a truly international response to the problems experienced in other regions. A joint response fosters a sense of cooperation rather than superpower meddling, and reduces the individual costs of involvement. Regional organizations, due to their background knowledge of a problem and established contacts with the conflicting parties, may prove particularly useful in dealing with the new variety of security threats. The US should seek out those foreign governments and international institutions which might best contribute to the resolution of a specific problem. Non-governmental organizations (eg International Red Cross) might also be helpful in resolving problems that could lead to instability *before* such conditions erupt.

Each of the problem areas cited in the previous examples of national security factors would be a candidate for concerted international attention before the resulting troubles could explode onto the world scene. We can either expend the effort – in time, energy and dollars – to address problems in the early stages, or we can wait until they turn into crises. We then have no choice but to spend far greater amounts of time and money – and often, put our soldiers' lives on the line – to restore order. The most expensive option is always to 'send in the troops.'

Practicing preventive defense

The final test of a workable preventive defense process will be based on whether the initiatives pursued make war less likely and create the conditions for peace. The process will need to be flexible and versatile in order to address a wide range of potential regional issues. Sometimes a model of a process used in another area might be adapted to the problem at hand with good results or perhaps a part of a negotiating framework might be used. Treaties and other international agreements, past and present, may hold lessons to apply to contemporary issues. In other cases, an entirely new approach may have to be created to address outstanding security problems.

Military matters

The *military* problem between North and South Korea affects the entire Northeast Asia region. The national interests of Japan, China, Russia, North and South Korea and the United States converge at the Korean peninsula, making it one of the most strategically important regions in the world. Tokyo suspects the Korean peninsula as a Chinese outpost. To Beijing, the peninsula historically has been considered as a potential Japanese land bridge to Manchuria. Since World War II, the United States has contributed directly to the security of the region by maintaining a military presence in South Korea, Japan and the surrounding waters. The Japanese now face the near-term prospect of a reunited Korea, possibly armed with nuclear weapons and backed by China. Japan's alignment with the United States may offer the best protective policy against such a threat. There are also distinct cultural clashes between the Chinese and Koreans on one hand, and the Japanese on the other. Resentment of Japan's dominance in the 1930s and 1940s runs deep among the Chinese and Korean people. The United States, perhaps in coordination with Russia, will be faced with a tough diplomatic task to ease the current military confrontation into a solid framework for regional stability as the new relationships in Northeast Asia evolve over time.

Political matters

The *political* issues in the Caucasus centered on Armenia and its future direction are focused on the prospect of sharing the wealth of oil in Central Asia by means of pipelines through the area. Through cooperation everyone wins, since stability is a precondition to enjoying the economic rewards of oil and its trans-shipment. Wise policies may spur cooperation where a win–win situation is created through cooperative relationships even while some national differences may remain. With the political issues between Armenia, Turkey and Azerbaijan so solidly entrenched in their perceptions of each other, a neutral observer group probably has the best chance to make progress on the differences that separate them. An international Contact Group similar to the one for the former Yugoslavia could have a very positive effect. A full-time mediator and staff could be required to resolve or manage differences in acceptable ways. Periodic meetings of the Contact Group could help the mediation process and give the three countries involved greater confidence in the process.

Economic matters

In the *economic* realm, Slovenia's success has been most impressive. In 1997, per capita income in Slovenia was $11,200 – higher than in Greece or Portugal. This was a far cry from Slovenia's condition in 1991, when Yugoslavia collapsed. The small country had lost 70 per cent of its markets, and its GDP had tumbled 23 per cent between 1987 and 1992. The Slovenes faced raging inflation and early conflict over privatization. The government quickly took action to reverse the GDP decline, bring inflation under control, and find new markets, quite successfully. Having been spared the economic chaos that afflicted other Central European countries, Slovenia has not had to pass through the political rigors of time-consuming tugs-of-war between the executive and legislative branches that others have experienced. Slovenia elected its first noncommunist Prime Minister in 1992, and the 1996 election resulted in a working coalition of parties that formed a new government. The Slovene authorities have never been accused of corrupting the election process as has been the case in some other post-Soviet Central European countries (Ramet, 1998, 113–18).

Perhaps a team of officials from Slovenia, joined by international experts from the main financial institutions, may offer Albania the best advice and assistance in rebuilding its economy, ravaged first by centralized planning and then again by the more recent unwise political actions.

Cultural matters

Similarly, a joint Castilian–Catalan team from Spain may be most helpful in Macedonia in dealing with issues of *culture* associated with the Albanian minority. The problems facing the Macedonians and the large Albanian population in the western part of the country are very similar to those experienced in Spain from 1939 until the late 1970s. The fascist regime of Francisco Franco attempted to quash and destroy the Catalan nation that had defied his takeover of Spain during the 1936–39 Spanish Civil War. For the next three decades, the Catalan language was banned in the conduct of government affairs. The Catalan language could not be taught in schools. Names of towns and primary streets in Barcelona were changed from Catalan to Castilian. Flying the centuries old Catalan flag was not permitted. Catalan festivals, plays and dances were banned. (Millas-Estany, 1998).

A robust democracy in the wake of the long dictatorship of Francisco Franco has developed a working relationship between the Catalan nation of about three million people and the Castilian-dominated government. Cultural and political freedoms are exercised by the Catalans within the context of Spanish sovereignty.

Demographic matters

The *demographic* problem facing many countries is rising population growth, which contributes to an influx of people from rural areas to cities, poor education, and unemployment. This concentration of unemployed, uneducated youths forms an ideal 'labor pool' for activist movements and radical organizations. The problem is structural. In the Middle East, for instance, accelerating population growth rates expand the labor force that in turn diverts government policy from economic investment to social investment. The result is a slower economic growth rate, and income distribution worsens as ratios of skilled to unskilled labor, and capital to labor, decrease. Investment capital is diverted to social infrastructure for sewage systems, water systems, and housing. These amenities of urban life are attractive to many, prompting even more people to move to the cities.

Countries trapped in the vicious cycle triggered by rapid population growth need international help to create jobs and provide public services. Bilateral aid, assistance from international institutions, and help from non-governmental organizations can ameliorate some of the negative consequences of growth and contribute to building the causes for peace (Stockton, 1996).

Moving forward

'The political object is the goal, war is the means of reaching it, and means can never be considered in isolation from their purpose' (Clausewitz, 1976, 87). The Cold War arms control process focused on the instruments of war, which were only a means to an end in the East–West struggle. The genius of the Marshall Plan was its focus on the causes of conflict, rather than the symptoms or means. The challenge for the United States and its allies at the threshold of the twenty-first century is not to repeat the Marshall Plan, but to take its philosophy and build on it by addressing the *causes* of conflict. Creating the conditions for peace based on the tenets of 'preventive defense' can enable the United States and its allies to deal more effectively with threats to regional stability.

The post-Cold War era is very different from the US-Soviet global standoff. We are faced with unfamiliar problems and new challenges. Speaking with one voice, the United States needs to first establish its own preventive defense process and then broaden it to encompass a variety of international partners. Outside actors – the UN regional organizations, coalitions of concerned countries, and non-governmental organizations – must work together in addressing the causes of instability. Development aid that promotes democracy, education and economic reform and growth is essential. We must also recognize that 'the legitimacy of outside assistance is important to its success. . . . Are states prepared to accept the legitimacy of such outside help?' (Holl, 1997, 14–15). Are we up to the problem-solving task required in this more complex world? Are the recipient countries prepared?

The international community has a choice. 'Two roads diverge in a yellow wood' – we can go down the well-trodden road that represents the Cold War security approach, or we can take the road 'less traveled by' (Frost, 1962). Taking the second road means adopting a preventive defense approach and building the conditions for peace in troubled countries. Which pathway shall we take?

Notes

1 The concept of international political instability is probabilistic. Stability can be considered from the perspective of the total system and the individual states that comprise it: (1) Systemic-Stability is '... the probability that the system retains all of its essential characteristics; that no single nation becomes dominant; that most of its members continue to survive; and that large-scale war does not occur', and (2) National-Stability is '... the probability of their continued political independence and territorial integrity without any significant probability of becoming engaged in a "war for survival."'
2 'The siting of mutually supporting defense positions designed to absorb and progressively weaken attack, prevent initial observations of the whole position by the enemy, and to allow the commander to maneuver his reserve.' *Department of Defense Dictionary of Military and Associated Terms.* Joint Publication 1–02, Joint Doctrine Division, Joint Staff.
3 In the spring of 1994, the United States, Russia, Britain, France and Germany established a five-nation Contact Group, with the goal of brokering settlement between the Federation of Bosnia and Herzegovina and the Bosnia Serbs. The United States and other Contact Group members convened two meetings and met with the Foreign Ministers of Bosnia, Croatia, and Serbia (also representing the Bosnian Serbs) in Geneva and New York during September 1995. The parties met in Dayton, Ohio, to begin 'proximity talks' on November 1. On November 21, 1995, the parties initialed the Dayton Peace Agreement. See US, Department of State, 'Bosnia Fact Sheet'.
4 This chapter was written long before the Kosovo crisis of 1999.
5 One of the Dayton Peace Agreement's central promises made by the international Contact Group and the Bosnia Federation, Croatia and Serbia was to return the refugees to their homes.

3
Integrating Environmental Factors into Conventional Security

Richard A. Matthew

Introduction

In the past two hundred years, as the environmental effects of industrialization have been felt in different parts of the world, the complex relationships between ecological and social systems have been prominent on the research agendas of many academic disciplines. It is only in the past 35 years, however, that environmentalism has gained urgency and international recognition, buttressed by scientific evidence of the multiple and dangerous ways in which humans are transforming their life support systems.

As evidence of the magnitude and complexity of environmental change has mounted, its implications for national and regional security have received greater attention. Since the end of the Cold War, academic and policy activity guided by concerns about the relationship between 'environmental change' and 'security' has mushroomed. In 1991 the concept of 'environmental security' was included in the US *National Security Strategy*. Five years later, in a speech that garnered global attention, former Secretary of State Warren Christopher stated: 'Environmental initiatives can be important, low-cost, high-impact tools in promoting our national security interests.'

Christopher's claim reflects the widely held belief that environmental factors are relevant to national security; it also displays the reigning lack of clarity about the nature of this linkage. His speech – like many official statements and documents – is general and vague, suggesting only that some environmental problems may be related to instability, conflict and other threats to national interests.

Perhaps any effort to lessen human-generated environmental degradation and scarcity has the potential to enhance human welfare and security; and perhaps some efforts will prove to be 'low-cost, high-impact'. But is it possible to move beyond such vague claims and aspirations and link the terms 'environment' and 'security' in ways that can serve as a more useful guide to research and policy?

In this chapter, I examine three approaches to making this linkage and argue that each is illuminating and important. I then consider in greater detail the approach that is most relevant to the official security community. I conclude that environmental factors can and should be integrated into traditional security affairs in two ways, and I assess efforts, opportunities and obstacles to doing so.

Three approaches to linking environment and security

It is a mainstay of Western philosophy that humans differ from other species in their capacity to reason. Reason allows us to identify goals, imagine and assess different ways of realizing these goals, and choose and implement the strategy that is most efficient. Among other things, our capacity to reason has enabled us to produce an extraordinary kit of tools, or technologies, designed to help us to reach our goals.

Many of these technologies have been developed to transform nature or insulate us from its inconveniences. For some observers, human history is the marvelous story of our successful efforts to increase our numbers, improve our health, lengthen our lives, and expand our comforts. But, environmentalists note, human behavior – from Easter Island to Mayan Mexico to the former USSR – often has pushed ecosystems beyond critical thresholds, leading to scarcity, decline and suffering (eg Feshbach, 1995; Ponting, 1991). Today our numbers and our technologies are having an unprecedented adverse impact on air, water, land and biodiversity (UNEP, 1997). Pessimists contend that human actions are generating an environmental crisis. The early warning signs – water scarcity, climate change, ozone depletion, biodiversity loss, air pollution – are clear, and we must change our behavior quickly and fundamentally (Wilson, 1992).

The optimists counter that humans can be – and have been – thoughtless and excessive, but overall our species is faring well – there are more of us than ever before living longer and healthier lives (Simon, 1989). There is no reason to believe that nature cannot tolerate whatever we do. Things change, but this has been the fate of the natural world for four billion years.

Between these two extreme positions lie numerous moderate views. One reason for this diverse intellectual landscape is that assessing our impact on the environment and the implications of this for ourselves and nature are not simply straightforward empirical matters. Our assessments are influenced by our values. For example, if wilderness is seen as a core element of American identity – something Americans will act aggressively to protect – then the US has an environmental problem because there is very little wilderness left – a few patches in Alaska, Montana and Wyoming. If our objective is to have all of humankind enjoy a Western style of living, then we have another sort of problem because the ways in which this might be achieved are likely to be highly destructive to many ecosystems.

In short, the significance of human-generated environmental change depends not only on what we require to survive and what nature can tolerate, but also on what we value and aspire to. While there may be some values to which all humans subscribe, many are widely contested. This adds an important social or political dimension to discussions of environmental change. It also helps explain a principal motivation behind exploring links between environmental change and security: security is widely perceived as a unifying value. The language of security is a proven way of mobilizing people and other resources to pursue a shared goal, and of trumping other values. Unfortunately, making this linkage has not been a simple task.

Two considerations help clarify the problems involved in trying to establish clear linkages between 'environment' and 'security'.

First, individuals and groups have attempted to link the terms for two fundamentally different reasons: the first is rhetorical (or political) and the second is analytical. Ironically, two rather incompatible groups have been politically motivated: environmentalists seeking to gain attention and divert resources to environmental issues by adopting the politically powerful language of security, and members of the security community seeking to protect budgets and identify new threats by tapping into the politically powerful language of environmentalism.

At the same time others (including members of both of the above groups) have explored linkages for analytical reasons, that is, in an effort to describe, explain and predict the ways in which environmental change might affect security and vice-versa. Rhetorical and analytical efforts have very different objectives: the former begin with the conviction that environmental change threatens human

Table 3.1 Perspectives on environment and security

Ecologist	Humanist	Statist
Preservation of nature (see Sessions, 1995)	Individual human welfare (see Myers, 1993)	Conventional national security (see Homer-Dixon, 1994)

welfare and seek to mobilize resources and build coalitions to respond to this threat; the latter seek empirical and theoretical evidence and justification for intuitions and fears. Unfortunately, the two objectives are often interwoven, resulting in a fair amount of confusion.

Second, the conceptual landscape is further complicated by the fact that 'environment' and 'security' are relatively elastic concepts that can be given very narrow or very comprehensive definitions. Because of this, it is not difficult to establish or challenge linkages between the two terms. For example, defining security in the very broad terms of individual human welfare promotes what I call a 'humanist' conception of the linkage between environment and security, whereas a traditional understanding that relates security to the core values of states (territory, people, culture) leads to a much more moderate – or 'statist' – stance. Although the field of debate has grown quite large, the predominant positions can be situated along a continuum as shown in Table 3.1.[1]

In assessing these three perspectives, it would be imprudent to understate the rhetorical value of linking 'environment' and 'security'. To some extent, politics is about managing perceptions, and insofar as this is true, whatever definition gains broad acceptance may prove to be a significant force in policy making even if it is not well-supported by scholarly research. Today the traditional security community appears to be winning the rhetorical contest, which is encouraging some environmentalists to abandon this strategy for gaining public attention and resources (Matthew, 1996).

On the analytical front, many heated debates are being waged. Ecologists challenge anthropocentric thinking and seek to view *homo sapiens* as one of millions of species that, together with inanimate material, make up nature. Humankind is an especially destructive species, and must be diverted from the brutal and damaging trajectory of its history. By 'following nature', in the sense intended by classical natural law thinkers such as Zeno and Chrysippus, that is, by adapting ourselves to natural patterns, rhythms and thresholds, we will not only cease those activities that are destroying our own life support system, but we may also recover some of the rich purpose of life that has been lost in our consumer society – spirituality, beauty, truth, simplicity.

Humanists, in contrast, are unapologetically anthropocentric. They tend to see a close relationship between the productive technologies that have exploited and degraded nature, and the economic, political and cultural structures that have degraded and exploited large segments of humanity. Awareness of environmental change gives us an opportunity – and a motivation – to rethink our relationship with nature and other humans. Humanists, following the example of human rights arguments, advise us to begin with the individual and his or her material and moral needs, and use this as a basis for criticizing current practices and designing new ones. Humanists fear a world in which we save the environment at the expense of the earth's poor and powerless; for them, we must harmonize nature and civilization, and seize this opportunity to correct imbalances that have appeared in all of our relationships.

Statists are motivated to consider the implications of environmental change within the framework of national interest and security. How do we protect our access to environmental goods beyond our border? How do we protect ourselves from externalities? How do we respond to crises generated or intensified by environmental degradation and scarcity – civil violence, transboundary population flows, regional instability and perhaps even interstate war? How do we predict when and where environmental change will create a situation in which we may have to use force?

In keeping with the Socratic character of our culture, lively debate is desirable insofar as it compels thinkers to clarify, modify and hence improve their arguments. It is less desirable, however, when it creates the impression that the range of perspectives constitutes a set of alternatives, only one of which can be correct.

And yet clearly there is an element of truth to this impression – on some issues, one cannot reconcile the various positions on environment and security. This might simply be the bane of reality – it is more like a collage of patterns, trends and preponderances than a structure of clear and distinct categories, and thus we have to live with contradiction, complexity and uncertainty. How else can one be, say, an ardent supporter of the market or merit or human rights and still retain a preference for the condition of friends and family?

But in large measure, the various perspectives on linking environment and security reflect different, rather than alternative, values and aspirations. The ecologist perspective expresses a fear of what Bill McKibben has called 'the end of nature' (1989). Whereas deep ecologists have made their bed, many others may value both nature – as a good in itself that ought to be preserved for aesthetic, moral or material reasons – and human society – as something equally valuable that exists in a relationship with nature that is difficult to specify. Even an individual concerned primarily with the status of human society might prefer to find some ground for reconciling this with a concern for nature. Similarly, one might be attracted to both statist and humanist perspectives, just as one might support the concept of a universal doctrine of human rights and still feel special attachments and obligations to friends and family.

In other words, only a few people are likely to choose the condition of the environment, or the condition of humankind, or the condition of the state, as the single and unconditional reference point for all reasoning and action. We are more complex than that, and more appreciative of the simultaneous separateness and interconnectedness of things that make up our world. Thus we understand the reality of compromise, tough decisions, changes of heart and dirty hands.

Indeed for those who might like to make a choice and run with it, the simple fact that social systems do not correspond with ecosystems constitutes a serious obstacle. To make an unconditional choice, one has to deny the immense power of the state, or the special status of the human species, or the transnational character of nature – denials that are not easy to sell or live by.

Awareness of the legitimacy of, and tensions among, the three positions noted above adds some complexity to the task of integrating environmental factors into national security thinking: where does the integration begin or end? But awareness also puts this project into a useful context by reminding us that: (a) environmental change is far more than just a traditional security issue; (b) traditional security assets may be of limited use in the grand scheme of things; and (c) involving the military in policy formulation and implementation comes with certain risks because, as Deudney (1990) has argued, the military world view is quite fixed and does not readily incorporate new sensibilities, which may be an important part of the long-term project of environmental reconciliation. Nonetheless, I will argue in the following section that there are ways in which environmental change is relevant to a traditional understanding of security, as well as to the institutions that exist to provide this good.

Integrating environmental factors into national security

There are two general ways in which one might seek to integrate environmental factors into national security thinking and practice. The first asks a question that is familiar to defense thinking, but has perhaps greater significance today than in the past: in what ways might environmental change threaten national interests and hence become relevant to the conventional mandates of military and intelligence institutions? The second way asks a more contemporary question: how do our national security practices affect the environment, and can they be modified to do less harm or even some good?

Insofar as the first question is concerned, three areas have received attention: (a) using security assets to protect access to environmental goods outside our borders, such as Middle East oil or international fisheries; (b) using force to respond to humanitarian crises generated, at least in part, by environmental problems, as in Somalia; and (c) bringing security assets to bear on situations in which environmental change has triggered, caused or amplified violent conflict in areas important to us or our allies (the conflict between Israel and its neighbors over water in the Jordan River Basin exemplifies this case; see Lowi, this volume).

In general, militaries are accustomed to protecting access to resources. Each year, for example, a score or more of conflicts related to over-fishing erupt in national or international waters. Some, such as the Canada–Spain dispute of 1995, lead to shots being exchanged. The important question here is, is this situation likely to escalate in the future such that national security planners ought to be preparing for it?

With regard to (b) and (c), the general problem can be depicted as shown in Table 3.2:

Table 3.2 Social responses to environmental change

	Adaptation	Ruination
Peaceful	1	2
Violent	3	4

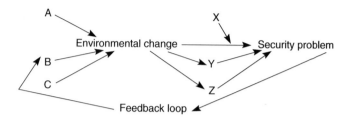

Figure 3.1 Causal model of environmental change and security

Casual observation and common sense suggest that when social systems are confronted with certain forms of environmental change such as resource scarcity or degradation, natural disasters, or epidemics, they adapt or fail to adapt and they do so peacefully or violently. Actual cases, of course, may display all four characteristics. The analytical challenge is to determine the factors conducive to each of the four outcomes noted in Table 3.2. The diplomatic policy challenge is to intervene so as to encourage more outcome 1 cases and fewer outcomes 4 cases. The security policy challenge is to be prepared to handle outcome 3 and 4 cases.

Much of the research done to date, notably through the three projects directed by Thomas Homer-Dixon, has focused on outcomes 3 and 4 cases, and as such has attracted the attention of the security community. But attempts to develop causal models that clearly demonstrate the role of environmental factors in triggering, intensifying or generating violent conflict or other severe security problems have been unsuccessful, although much has been learned about the relationship. Typically, analysts have produced models along the lines shown in Figure 3.1 (see Homer-Dixon, 1994 and 1996):

In this model, A, B and C refer to the anthropogenic causes of environmental change. Homer-Dixon (1994, 1996) defines these as increases in demand on resources due to population growth or development; decreases in supply due to resource degradation or depletion; and changes in distribution due to poverty or some other form of access denial. X, Y and Z refer to the factors widely regarded as necessary, sufficient or enabling causes of security problems (eg weak political institutions and corruption; poverty and/or inequity; activist military and accessibility of arms).

In such causal networks, weighting environmental factors has proven very difficult. There is little doubt that they are a part of

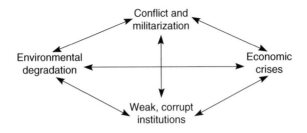

Figure 3.2 Development trap (adapted from Mohammed 1996, 2)[2]

the network of variables that generate insecurity within and between states. The conditions under which they play a key or decisive role is far less clear. However, this is true of most models that have sought to describe, explain and predict conflict and instability.

It is unlikely that we will be able to improve our causal models until (a) substantial research has been done on cases in outcomes 1 and 2 of Table 3.2, and (b) the work done on cases in outcomes 3 and 4 is connected more systematically to the available literature on conflict and insecurity. Moreover, scholarly work needs to be assessed and refined in light of the sort of hands-on experience that sustains considered judgments. This is because political outcomes are the product of general principles, context and contingency. For this reason, I believe that case study or regional analysis by teams of scholars and practitioners is likely to be the most fruitful approach to understanding this particular linkage between environmental change and security.

There are good reasons to undertake such research. For example, several recent studies suggest that the poorer countries of Africa constitute an area in which environmental factors may play an important role in triggering or intensifying conflict (eg Kaplan, 1994). Nadir Mohammed (1994, 1996) has modeled what he terms a 'development trap', in which conflict and militarization, environmental degradation and economic crisis interact in a mutually reinforcing and inescapable manner (Figure 3.2).

Mohammed foresees direct military conflict over scarce resources in the near future, especially in the fifteen countries with higher than average population density.[3] (1996, 5 and 10) He also predicts that environmental degradation will cause population displacement and economic crises that will in turn generate conflict and violence (1996, 5).

Level of interest in area

		High	Low
Capacity to affect outcomes	High	1	2
	Low	3	4

Figure 3.3 Motivation/capability grid

Mohammed's model does not make a clear distinction between civil and interstate conflict. Homer-Dixon and others have argued persuasively that environmental problems, especially scarcity, exacerbate civil strife in developing countries and weaken public institutions but rarely lead to interstate war. With regard to water, the work of Aaron Wolf provides support to this claim, although Wolf (1997) acknowledges that the past may not predict the future. Homer-Dixon (1996) suggests that these findings should be of interest to policy makers because prolonged instability and civil strife in the developing world can have an adverse impact on our needs and objectives. This suggests another problem that must be answered in order to build a bridge between analysis and policy. The problem can be depicted as shown in Figure 3.3.

What criteria should be used to make prudent assessments of whether a country has (a) an interest, and (b) the ability to intervene effectively in areas where environmental factors are generating or exacerbating conflict and violence?

If we keep in mind that UNEP's first *Global Environmental Outlook* report (1997) concluded that environmental degradation and scarcity are generally worsening worldwide, then the arguments of Homer-Dixon, Kaplan, Mohammed and others seem increasingly important. At the same time, the connections between environmental change and the sort of threats to national interests that require the use of force tend to be either self-evident or buried in a dense causal network. Further work, then, is required to enable security specialists to anticipate this type of threat. At the time of this writing, at least three major research efforts are underway to fill this need. The CIA's new Environmental Center is seeking to identify critical thresholds beyond which conflict is likely; Pacific

Northwest National Laboratories' Environmental Security Center is developing a method of modeling environmental threats; and the IHDP's Global Environmental Change and Human Security Project is refining its Human Vulnerability Index, a measure that aggregates economic, health, political and environmental indicators in an effort to strengthen predictive capacity. Based on research done to date on the positive relationship between democracy and lower levels of violence (Rummel 1997), it would seem wise to examine the connection, suggested by Homer-Dixon, between environmental change and authoritarianism.

Clarifying the conditions under which environmental change might pose a threat to national security may prove to be an inconclusive enterprise, although much about the relationship is being learned in the process. While our analytical and predictive skills are likely to remain imperfect, there is nonetheless a need for security institutions to develop response mechanisms. In the United States, for example, the concept of 'preventive defense', as discussed by Mary Margaret Evans in Chapter 2, has emerged in the 1990s; as Sherri Goodman, Deputy Undersecretary of Defense for Environmental Security, explained in a 1996 speech, in line with this concept her objective is 'to understand where and under what circumstances environmental degradation and scarcity may contribute to instability and conflict, and to address these conditions early enough to make a difference' (quoted in Matthew 1996, 43). Already, significant efforts to do this have been orchestrated by the militaries of the US and Australia, as well as NATO. Numerous meetings have been held to encourage dialogue on environmental issues as a challenge to security, and various bilateral and multilateral agreements have been signed (eg Canada/US/Australia; US/Russia; US/Sweden) to manage certain environmental problems that might otherwise be a source of tension and conflict (Matthew 1997, 1998).

In summarizing the state of our answers to the first set of questions about the relevance of environmental factors to national security, the following propositions are credible, if incomplete:

1 In the widely remarked shift in violence from interstate to intrastate forms, environmental degradation and scarcity are almost certainly contributing variables. Given the continuing escalation of environmental problems, this trend is likely to grow. Resolving such violence will require sensitivity to environmental factors and, usually, multilateral cooperation in view of the transnational character of environmental problems.

2 Although historical evidence suggests environmental factors rarely figure prominently among the causes of interstate war, the increasing magnitude of environmental change suggests the future may deviate from the past.
3 R.J. Rummel's (1997) comprehensive analysis of the strong positive relationship between non-democratic regimes and all forms of violence (including mass social violence, state sponsored violence and interstate violence) suggests that if, as Homer-Dixon has argued, environmental problems strengthen non-democratic politics, then we should be concerned about the long-term prospects of environmental change creating political conditions conducive to violent conflict.

We are on firmer ground in answering the second set of questions. Recent studies, confirmed by an expansion of policy activity, indicate there are several ways in which the security community can contribute to lessening environmental degradation and scarcity:

A. Militaries have a well-deserved reputation as major sources of pollution. By greening military training, testing and warfighting activities, substantial gains can be made. In an effort to improve its environmental record, for example, the US military claims to have reduced toxic waste by 50 per cent in the 1990s (Butts, 1994). It is unclear, however, to what extent this reduction is due to defense downsizing. Still, the US military has adopted a general policy of compliance with environmental regulations; in joint venture with Sweden, it has developed a set of guidelines for military operations; it has cooperated with Russia to reduce radioactive contamination in the Arctic; it has developed new recycling technologies; and it has engaged in environmental restoration projects, notably in the Chesapeake Bay area.

B. The sophisticated data gathering and analysis capabilities of the intelligence community offer another opportunity for making a contribution to environmental policy initiatives. In 1992 Vice President Al Gore established the Medea Group, pairing 70 civilian scientists with 70 intelligence analysts to assess the value of archived satellite imagery. At the same time the National Intelligence Council was set up to find ways of making intelligence available to new consumers. To date, US intelligence has helped UN agencies and others in assessing deforestation, monitoring population flows, and tracking forest fires (Matthew, 1996, 1997). There continue to be concerns, however, about the reliability of this data (Deibert, 1996) and anecdotal evidence suggests access

to it remains sporadic and difficult. The CIA's new Environmental Center may prove to be a successful conduit for intelligence, but it is too early to evaluate its efforts.
C. At least in principle, the security community could be mobilized to support a wide range of environmental projects. As Kent Butts (1994) has argued, the US Corps of Engineers possesses skills that could be directed to restoration projects. A UN force might strengthen the world's capacity to monitor and enforce compliance with multilateral environmental agreements (MEAs) negotiated in the past three decades.
D. Finally, security communities do constitute a form of transnational association that can be used to disseminate information, transfer technology, and promote dialogue on environmental issues. Under the auspices of the Partnership for Peace Program, NATO's Science Committee has sponsored regular Advanced Research Workshops, enabling military and civilian specialists from NATO and non-NATO countries to meet on an extended basis to discuss environmental issues of common concern. The Asia-Pacific Center for Security Studies, part of the US Department of Defense, has hosted high level meetings on environmental change and regional security involving senior officials from throughout the area. The Army War College has organized environmental security simulations and meetings. These initiatives are illustrative of extensive behavior designed to promote transnational dialogue, transparency and cooperation.

These suggestions and initiatives to green the security community and make some of its resources available for environmental projects are likely to evoke skepticism among those who regard military and intelligence organizations as secretive, unreliable and committed to protecting national interests at any cost. It would be premature, however, to offer an assessment of activities that are still at an early stage of development. It is not premature, however, to ask whether we should be encouraged or dismayed by the flurry of activity witnessed since the end of the Cold War.

Is the traditional security community, prone to secrecy and primed to use force, using its extensive resources to colonize and reshape the arena of environmental policy, or has environmental awareness finally penetrated this previously unreceptive arm of the state? The great risk, perhaps, is that the fears expressed by Homer-Dixon and Robert Kaplan, are gradually becoming a reality. The magnitude of environmental change may be creating conditions ripe for the use of force and the revitalization of authoritarian governance. Military and intelligence specialists are positioning themselves – gaining expertise, information, resources and trust – to take over, if necessary, environmental policy by redescribing it as a security issue.

Or, rather than accept this pessimistic realist analysis, one might argue that the security community is beginning to accept that it exists in a complex and interdependent world in which, as Robert Keohane and Joseph Nye (1974) have argued, brute force is of limited utility for addressing many pressing problems. The security community must either adapt to this reality, sharing its expertise and resources, or risk budget cuts.

I would suggest that the integration of environmental concerns into traditional security thinking and practice falls between these two perspectives: force may be required to address some environmental problems, or the consequences of these, but security assets can also be used to support environmental projects in other ways. What is essential is that societies do not come to rely upon their security institutions to address environmental problems and their consequences. Societies must encourage the integration of environmental factors into national security affairs while remaining aware of the risks and limitations of this approach.

Conclusions

There are areas in which national interests may be threatened by environmental change such that the use of force will be necessary. There are areas in which environmental change may produce or intensify humanitarian crises that will require some application of force to resolve. And there are areas in which the security community can contribute to environmental initiatives by virtue of its extensive resources and sophisticated technologies.

But this is new terrain for the security community and parts of it are loathe to undertake activities that might diminish or even compromise their primary national defense mission. Perhaps more important, other groups – including foreign states and environmentalists – are worried about the broader implications of involving the security community in environmental policy.

Efforts to integrate environmental factors into national security thinking and practice are hampered in several other ways. Our understanding of the relationship between environmental change and conflict is incomplete. Countries such as the US, in which these efforts are most advanced, do not yet have clear environmental foreign policy agendas, which makes it difficult, perhaps impossible, for different agencies to coordinate their efforts, or mobilize around clear goals. Interagency rivalry for prestige and resources further undermines this integration.

In view of this it is important to be reasonable about the limitations of what can be achieved by linking environmental change to national security. Under certain very specific conditions, the linkage is viable and useful. But in large measure, environmental change does not pose a conventional security threat. It is a complex, long-term phenomenon that affects social systems in a variety of ways. While it may contribute to conflict in areas where policy options are already severely limited, elsewhere it is likely to stimulate technological innovation, promote interstate cooperation, and lead to behavioral changes at the micro level. If such response and adaptation mechanisms fail, it may well lead to a gradual reduction in the quality of human life. In either case, it does not pose a traditional security threat.

Similarly, the value of security assets is also limited. The roots of environmental change lie in poverty, population growth and distribution, consumption rates, production technologies and waste management practices. To address environmental change effectively is likely to require attacking some of its root causes. To this end, education and research, international data sharing and regimes, partnerships that include public and private actors, and sustainable development initiatives may be the most important elements of an environmental rescue strategy.

In conclusion, I believe there are fruitful ways of linking environmental change to a conventional understanding of security, and that security assets may be applied to addressing some environmental problems. Focusing on this set of issues will reinforce initiatives begun in the last few years and the benefits of such an approach are likely to be realized quickly and at relatively low cost. To ignore the role of environmental change in fostering conflict and regional instability would be foolish, even if it is rarely the primary cause of such problems. To avoid using military assets for environmental ends perpetuates a 'we versus they' mentality that is at odds with the pervasive, inclusive nature of environmental problems, and with the need to address them with all available tools. At the same time, linking environmental change to national security does not provide a comprehensive framework for analysing and addressing environmental problems, the bulk of which do and should lie outside this framework.

Notes

1 One could imagine a fourth position – rejectionist – comprised of writers such as Daniel Deudney (1990) and Marc Levy (1995b) who believe that environmental security is a misleading, foolish or dangerous concept. These scholars are critical of the research that has been undertaken on the relationship between environmental change and conflict, and worried about the implications of involving the military in environmental issues. They are, at heart, highly skeptical of the statist approach to linking the terms, and concerned that, by using the loaded language of security, other approaches may bolster – albeit indirectly – the statist approach. But rather than present this as a fourth category, the null category, these arguments will be considered later in the chapter.
2 To this model I have added the category of weak and corrupt institutions.
3 These are: Benin, Burundi, Egypt, Ethiopia, Gambia, Ghana, Kenya, Lesotho, Malawi, Morocco, Nigeria, Rwanda, Swaziland, Tunisia and Uganda.

4
Security, Governance, and the Environment

Steve Rayner and Elizabeth L. Malone

In the animal kingdom, there is a major distinction between organisms that have hard outer shells or external skeletons and those that have their bony support structures on the inside. The hard-shell species rely for their security on the integrity of their covering. Thus they are immune from predators until the shell is broken, at which point they rapidly succumb. The price that hard-shell species pay for their initial immunity is limited mobility and inflexibility of response options when attacked. In contrast, soft bodied animals with internal skeletons are easily damaged by attackers but repair more rapidly. Furthermore, their higher mobility and greater flexibility enable them to avoid many hazardous situations and to respond adaptively when they are unavoidable.

Thus, the animal world provides a metaphor for two approaches to national and environmental security. The first approach is a hard-shell one that focuses on securing the borders of the state against attack or safeguarding the decision making apparatus of the state from subversive infiltration. The second approach focuses on developing the internal framework or skeleton of the state to maximize its capacity for resilience and adaptation in the face of political, military, and environmental challenges.

Unfortunately, neither zoology nor history seems to support the notion that one can maximize both strategies simultaneously. The more that hard protection is concentrated at the boundary of the organism or the state, the more it becomes committed to a limited territory. The greater reliance on internal structure, the less tolerance the organism or the state has for the heavy baggage of armor and limited territorial range. In the case of the state, the demands for free trade, open markets, labor mobility, and so on, seldom, if ever, emanate from militaristic regimes.

Thus the conventional use of the term 'national security' evokes images of barbed wire border fences and tank movements, or of government agents working clandestinely to prevent espionage, sabotage, or attack. The supposed anarchy among sovereign states gives rise to suspicion and distrust; each nation-state must be on guard at its boundaries lest another nation-state learn its secrets, or wreak social, political or military havoc. This is essentially a hard-shell approach to security.

But another strategy to build national security is possible. We suggest that the term 'security' be considered primarily as the capacity for *societal resilience*. National security, then, becomes an issue of governance – not just of government. Thus defined, security is the ability of the social system to resist the impact of a wide variety of disruptions, not just military attack or subversion. The term 'environmental security' also undergoes a transformation in this approach. Rather than focusing on environmental causes or consequences of military action, environmental security is extended to encompass the environmental impacts on stability *within* a country. These may be as important, sometimes more important, than cross-boundary environmental issues. In this regard, the relationship between environmental problems and local institutional arrangements is the key to dealing with the challenge of building a stable structure of governance.

The activities involved in the boundary setting kind of national security are rather specialized technical activities, with military overtones. Activities involved in the stabilizing kind of national security have broader relevance. For example, training and technical assistance activities can be thought of as building national and local capabilities for governance to manage environmental problems. Experience in negotiating workable agreements to halt pollution or clean up polluted areas is another example of a governance capability directly relevant to environmental and national security.

Our broad use of the term 'security' has a long and honorable history. John Stuart Mill (1806–73) strongly identifies happiness with security. Along with nutritional requirements, Mill claimed that security was the single universal human need. Mill's idea of happiness, like the standard definition of security, is defined by the absence of threat: it is the absence of pain, with occasional pleasures, but above all with the knowledge that one will not be arbitrarily dispossessed. To achieve happiness, people must know that life will be fair, that is, that those in authority are just. Governance, then, is a crucial factor in a citizen's happiness and security.

In this paper we espouse a soft-shell strategy. We argue that those who are interested in national and environmental security should invest in a wide variety of institutions, not for their immediate instrumental contribution to the technological effectiveness or economic efficiency of a project, but as ends in themselves that also ultimately contribute to long-term national and international societal, economic and ecological security. We will build our case on recent research into the fundamental prerequisites for a sustainable society: research by Robert Putnam (1993) into the correlation between associational activities, and economic and governmental success; and Francis Fukuyama's (1996) analysis of high- and low-trust societies and their respective abilities to build well-functioning democracies.

We take the argument further than either Putnam or Fukuyama, though, in suggesting that the relevant investment for long-term national and environmental security is in a wide diversity of social activities that, at first blush, appear to have little, if any, relevance to providing basic human needs, national infrastructure, or technical capacity building. Examples of such activities include, but are not limited to, sports clubs, carnival societies, cultural associations, hobbyist clubs (such as gardening and photography clubs), professional societies, craft guilds, and scientific, musical and theater groups, as well as community associations and resource-user groups. The Tanzanian 'secret dancing societies' discussed by Cernea (1993) would be another example. The important points are to assume diversity and decouple support of associations from their immediate usefulness as channels for military, technological and economic development, thus following the advice of the celebrated English landscape architect Lancelot 'Capability' Brown (1716–83) to 'confront the object and draw nigh obliquely'.

We suggest that the merit in these tangled associational networks is that they provide societies with plural viewpoints to draw on in changing conditions. That is, the society can switch strategies when its predominant approach is no longer working. For example, a society could replace incentives with regulations when the former become ineffective, or allow successful individuals to lead the way in introducing innovations when the usual procedures fail.

Investing in social activities will provide multiple, overlapping links among people which will build society by supplementing other ties that are traditionally used to measure stable governments: ethnicity and civic nationalism (Ignatieff 1993). Ethnicity, as a basis for forming a national government, claims that a person's deepest attachments are inherited. Civic nationalism bases governmental ties on a chosen common political creed rather than on 'blood' ties. That ethnic and civic ties exist is incontestable; but of themselves they do not guarantee stable governments. A commitment to the same political creed can be undermined by ethnic strife; Canada and Northern Ireland provide examples. Multiple ethnic backgrounds divide the former Yugoslavia but exist amicably in Switzerland under a strong government.

Something more than ethnic or civic nationalism must be at work in a truly sustainable society, and we suggest that something is the disorderly network of social ties that make people value and work towards maintaining a vigorous and stable government.[1] This idea is consistent with a large literature, including the suggestion by Esman and Uphoff (1984) more than a decade ago that a vigorous network of participatory membership organizations is essential to overcoming mass poverty in the less industrialized world. It is consistent with Putnam's (1993) detailed empirical demonstration of the positive relationship between a rich and diverse civic life in Northern Italy, and the economic and governmental success of the northern provinces. And it is compatible with Fukuyama's (1996) more recent argument that civic plurality is essential to a well-functioning democracy.

The correlation of diverse associational ties and successful governance is well illustrated by Putnam's careful empirical analysis of the contrasting development patterns of government effectiveness among the Italian regions. Putnam and his colleagues constructed a multiattribute index to measure the institutional performance of each of Italy's 20 regions, using a heterogeneous list of development factors that includes the sorts of elements that would be familiar to most people (see Table 4.1). Moreover, subjective citizen satisfaction with government was measured, producing highly correlated results.

The highest performers overall are the former Northern Republics. The former Papal states of Central Italy perform less well, while the Southern Mezzogiorno region (formerly comprising the kingdom of Naples, and before that of Sicily) do least well. This pattern of institutional effectiveness is not surprising. Most people with

Table 4.1 Putnam's Index of Institutional Performance

Index of Institutional Performance 1978–1985: Performance Indicators (Source: Putnam, 1993)
Reform legislation
Day-care legislation
Housing and urban development
Statistical and information services
Legislative innovation
Cabinet stability
Family clinics
Bureaucratic responsiveness
Industrial policy instruments
Local health and spending
Agricultural spending capacity

even a passing acquaintance with Italy are aware of national self-consciousness concerning the 'European' North and the 'African' South. The common explanation for these differences looks reasonable but is unsatisfactory when examined closely. This explanation is geographically determinist – that, being closer to the European heartland, the North has shared in its economic modernization. But we are then left with a further question: why does the cut-off point for modernization fall where it does, halfway down the Italian isthmus? Why not further north (for example, at the Alps) or further south (for example, at the Straits of Messina)? Furthermore, the North–South disparity cannot be explained simply by the distribution of economic resources. In fact, funding for regional governments is provided by the central authorities according to a formula that favors backward regions. Also, economic resources cannot explain the distribution of successful governance within the North and South: (a) Campania is wealthier than Molise and Basilicata, but the latter's governments are more successful, by both objective and subjective criteria; (b) Lombardia, Piedmonte, and Liguria are economically much better off than Emilia–Romagna and Umbria, whose institutional performance scored higher.

Another possible correlation would be between educational levels of the citizens and institutional performance, but again Putnam finds that this is not the case. Educational levels of those involved in government are in fact higher in the South than in the North.

Geography, economic resources and education are not sufficient explanations for why northern regions have higher institutional performance ratings than southern regions. The factor that most closely correlates with institutional performance (and, incidentally, citizen satisfaction with government) is what Putnam terms *civic engagement*, as measured by the strength and vitality of associational life – membership in choral societies, amateur soccer clubs, hunters' associations, bird-watching clubs, etc. Independent associational life outside of the family and the structures of the state is a major component of the index of civic community that Putnam constructs for comparison with institutional performance of regional governments.

The high correlation of civic community and institutional performance is startling and compelling. Not only does the correlation distinguish the high performance regions from the low performers, it accounts for variations in performance *within* both high- and low-performing categories. The correlation is particularly striking when mapped geographically. Putnam's results also include a negative correlation of civic community with *clientelism* – dependence on vertical axes of dependence and loyalty – that characterizes political and social relationships in the Mezzogiorno, where the regional elite is drawn almost entirely from the most economically and educationally privileged portion of the population.

A major strength of Putnam's study is its longitudinal scope, which enables him to investigate the direction of causation between civic community and economic development. As he says,

> Both the North–South gap in Italy, and the range of theories that have been offered to account for it, mirror the broader debate about development in the Third World. Why do so many countries remain underdeveloped: inadequate resources? government mistakes? center–periphery *dependencia*? market failures? 'culture'? Precisely for that reason, studies of the Italian case have the potential to contribute importantly to our understanding of why many (but not all) Third World countries remain inextricably mired in poverty. (Putnam, 1993, 159)

Does a strong civic community result from economic development, or does the civic community create conditions for economic growth? Putnam's statistical analysis of these pathways (Figure 4.1) reveals an astonishing finding, well worth the attention of everyone concerned with national security. It turns out that the assumptions shared by both Marxists and free market economists

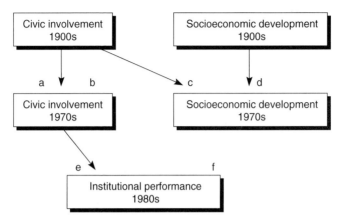

Figure 4.1 Actual effects among civic involvement, socioeconomic development, and institutional performance: Italy, 1900s–1980s (Putnam, 1993).

that the economy ultimately determines the shape of society is quite misplaced. Using measures such as agricultural and industrial employment, Putnam shows the following:

1. Civic involvement 1860–1920 is a very powerful predictor of contemporary civic community.
2. Economic development (industrialization and public health) has *no impact* whatsoever on contemporary civics.
3. Civic involvement 1860–1920 is a very powerful predictor of present socioeconomic development.

In other words, while a strong civic community reproduces itself *and* generates economic development (external conditions being favorable), economic modernization alone does not generate a strong civic community and, ultimately, becomes unstable. This is because, unlike financial capital, the social capital embodied in the civic community is what Hirschman (1984) terms a *moral resource* whose supply increases with use, rather than decreases, and which becomes depleted if not used.

Putnam's empirical work alone should demonstrate the folly of continuing to pour the overwhelming bulk of our development investment into economic and technological projects in which nourishing civic life is subjugated to the instrumental details of implementation, rather than treated as an end in itself. For example, if Putnam's view is correct, sectarian strife in Northern Ireland will never be eliminated by economic development, yet this has consistently been the basis of British Government policy in the province for over 25 years. But Putnam's results should not be a surprise. Adam Smith (1723–90) – hero of the free market economists – told us as much in his lesser-known work *The Theory of Moral Sentiments* (1759), where he describes the forgotten feet of civil society without which the hidden hand of the market is dismembered from the social body. Social capital, in the form of civic life, is a precondition of economic success. We are putting the cart before the horse when we tackle security as primarily an economic issue.

Here we should stop to comment about the difference between poverty and destitution, certainly a key distinction when we talk about security (Douglas, 1982). Famine, malnourishment, insufficient housing, and lack of clothing are what we usually associate with poverty. (Putnam seems to have this notion in mind in the quotation above.) But these are surely the markers of destitution, a crisis situation that is not only unsustainable but intolerable. Poverty, on the other hand, is a restriction of the choices available to people; the issue of not having goods is important because certain goods help people take their places in society, committing others to their activities. Poverty results from the systematic exclusion of people from society, and its effects are cumulative (for example, lack of education leading to low-level jobs throughout a lifetime). Poverty cannot be defined by level of income ('the poverty line').

Destitution and poverty, being different states of being, require different approaches. Destitution is an emergency situation, calling for immediate relief. It is sustainable neither in the short term nor the long term. Poverty is a social issue, and though we could argue that it may be a permanent feature of most societies (that is, most societies will systematically exclude one or more groups of people), surely the ability of all members of the society to participate in its civic life is a goal whose achievement would constitute social security.[2]

The issue of the relationship between poverty and security gives us a motive to look behind Putnam's empirical findings and ask the question, 'Why does civic community play such a key role in security?' This question is important for everyone. Attention to the development of social capital is not just an imperative for emerging industrial economies. Francis Fukuyama, in his book beguilingly entitled *Trust* (1996), argues that the extended moral network of civic society is a precondition for competitive success in the new global economy. Conversely, absence of civic community leads to fear of the state, accompanied by demands for a strong state to regulate the behavior of others who cannot be trusted (Banfield, 1958).

Fukuyama, Putnam and other scholars ascribe the efficacy of social capital – defined as a rich, overlapping pluralistic civic life – to its role in creating societal trust. It is certainly true that it is harder to dehumanize your political opponents or economic rivals if you play cricket with them on Sunday, go to mosque with them on Friday and attend a charity fund-raiser with them on Wednesday. It is also true that you are more likely to behave in morally prescribed ways if you have confidence that others will stick to the rules.

But part of the story is missing here, hidden under the labels of 'civic community' and 'trust'. Civic community and trust have different faces in different societies. Face-to-face communities, in which all members are known to each other are not necessarily 'high trust' societies; individuals in egalitarian groups, in which everyone has an equal claim to the truth, tend to trust themselves as much, if not more, than they trust others. Moreover, diverse forms of trust exist (Rayner and Cantor, 1987). And a truly robust civil society has the capacity to function even where trust is tenuous. So, without distracting from the importance of trust, we suggest that Putnam and Fukuyama are missing an important part of the story.

The importance of civic pluralism and the associated trust networks is that they imbue society with an enhanced capability for *complex strategy switching*. That is, the society will be able to draw on its plural viewpoints to respond to new situations in new ways. Security in its essence cannot be the pastoral fantasy of limiting development only to those things we can do over and over for ever and ever. It cannot be the hard-shell conviction that strong borders are enough protection. It must be built on the ability to switch from a strategy that is not working, or will not work for much longer, to one that will work, at least for now.

We return to Italy for an illustration of our assertion. Putnam's entire book is based on Italy, and Fukuyama devotes a chapter to the differences in trust among the North, the Mezzogiorno, and the Terza Italia in between. Although Putnam does not address complex strategy switching, he undergirds his statistical analyses of the past century with an historical account of the origin and resilience of civic community in Northern Italy that spans almost a millennium. Fukuyama offers a further distinction between societal trust that is based on family ties, and trust that includes nonfamilial ties among people.

The 12th century saw the emergence of two Italies. In the South, Norman conquerors combined feudal and Byzantine elements to produce a flourishing military, scientific, artistic and commercial center that was the richest, most advanced, and highly organized in Europe. The civic life of merchants and artisans was centrally regulated by the state. But as the kingdom declined, absolutist barons appropriated the roles of the state, establishing the basis of the system of clientelism that persists to this day.

In contrast, the city states of Northern Italy (Florence, Venice, Bologna, Genoa, Milan) established self-governing *communes*, coordinated by decentralized guilds, public-works associations, business partnerships and fraternities, which successfully absorbed the rural aristocracy of the region. The *communes* produced the first professionalized public administration, an independent judiciary for administering contracts, settling disputes, and so on. Most importantly for economic development, this provided conditions for the emergence of the first modern system of credit.

Despite plague, foreign invasion, and the appropriation of political power by autocratic families, the civic tradition of the republics never died. (Even the Medicis and Gonzagas had to rule with and through the civic institutions.) In the eighteenth century, the civic structures began to reassert their primacy over the autocratic once again.

Fukuyama builds on Putnam's analysis by identifying trust as the essential element in economic development. The North, with its persisting tradition of civic participation, extends trust networks into all kinds of social ties and therefore has large-scale economic firms. However, Italians in the South tend to trust each other within family circles but not outside those circles, with corresponding implications for the amount of economic activity outside those circles of trust. In the so called Terza Italia, between the South and the heavily industrial North, a high level of trust exists within family circles in family-owned business and in networks of such businesses. So, differences in the scale of economic organizations parallel the differences in types and extent of trust relationships. The North is characterized by 'civic-mindedness' and large businesses, the South by 'amoral familism' and small, weak firms.

But the explanation of these different development paths cannot stop at the simple assertion that interpersonal trust survived in the North but was never able to emerge in the South. Trust can be built within hierarchies, as well as among peers and within markets, though each form of trust has its peculiarities and particularities (Rayner and Cantor, 1987). Hierarchies trust rules and routines, egalitarian groups trust participatory processes, and markets trust successful individuals.

What the Southern kingdom of Sicily (later Naples) lacked, that the North possessed in abundance, was the availability of non-hierarchical strategies for living together. Lack of variety and demonstrated alternative ways of doing business condemned the South to stagnation. Lack of alternative models of trust to the patron–client relationship was indeed part of the institutional poverty of the South, while the coexistence of hierarchy with egalitarian and market social relationships constituted the social capital of the North.

The term *complex strategy switching* includes no fewer than three strategies for managing society and, by extension, managing natural resources: markets, hierarchies, and egalitarian peer groups or collectives. Both mathematical and social theoretical reasons direct our focus on these three strategies.

Three optional states is the minimum requirement for a complex system. Simple systems, such as dichotomies, cannot be dynamic. At best, they can only flip back and forth between one state of affairs and another. The system is either dominated by State A or State B. As such, they are not capable of sustainable adaptation. For example, the successful merchants of the twelfth-century kingdom of Sicily had access to the model of the market as a potential alternative to the centralized state. But without a readily available complementary model of egalitarian peer groups to organize themselves in defiance of hierarchy, they remained stuck within a static system.

Introducing a third element into the system increases the possible permutations of dominance among States A, B or C to 15. Furthermore, these can then be sequenced in many different ways, generating a variety of different development paths or strategies. If I seek to maintain my car with a toolkit containing only two or four tools, I am likely to have more difficulty keeping it running than if I have a larger bag of diverse tools. Similarly, a society drawing on plural social strategies will be more sustainable than a society with a very limited set.

Markets, hierarchies, and egalitarian collectives provide the minimum requisite social variety for a self-adaptive socioeconomic system and the minimum requisite variety for a resilient and dynamic civil society. We focus on these particular three elements because of their persistence in the social science literature.

Unfortunately, this persistence has been obscured by the tendency of social scientists to represent complex social realities by dichotomous variables (see Table 4.2). In the mid-nineteenth century, the legal historian Sir Henry Maine (1861) distinguished social solidarity based on *status*, in which actors know their place in hierarchical structures based on the idiom of the family, from solidarity based on *contract*, in which agents freely associate by negotiated agreement. Later in the century, German sociologist Ferdinand Toennies (1887) distinguished between *gemeinschaft*, where societies are bound by ties of kinship, friendship, and local tradition, from *gesellschaft* where social bonds were created by individualist competition and contract. At the turn of the century, French anthropologist Emile Durkheim (1893) distinguished human societies based on *mechanical solidarity*, in which agents bind themselves to others on the basis of sameness, from those built upon *organic solidarity*, in which agents are bound together by the interdependence of specialized social roles.

Each of these grand dichotomies was viewed by its author in evolutionary terms which continue to resonate in contemporary social theorizing, such as that of Bennett and Dahlberg (1990) who, echoing Durkheim, detect in the development from preindustrial to industrial society a shift from *multifunctionalism*, where everyone can do everything, towards *specialization*. For the most part, however, more recent approaches dispense with the unidirectional evolutionary assumption. Educational psychologists have identified *positional* families in which behavior is regulated by appeals to hierarchical authority and *personal* families in which behavior is regulated by appeals based on individual preferences (Bernstein, 1971). Major contemporary political scientists and economists such as Charles Lindblom (1977) and Oliver Williamson (1975) focus on the different characteristics and dynamics of coexisting and competing social systems based on the social bonds created through participation in *markets* and those based on the solidarity of *hierarchy*.

There is a great deal of overlap among these grand dichotomies of social theory. However, they are far from perfectly congruent and, in sum, give rise to three, rather than two, basic forms of social solidarity (Table 4.2).

Table 4.2 Characteristics of three kinds of social solidarity (as described in classic social science literature) (Rayner, 1995)

Market	Hierarchical	Egalitarian
Gesellschaft	*Gemeinschaft*	*Gemeinschaft*
Organic solidarity	Organic solidarity	mechanical solidarity
Specialized roles	Specialized roles	multi-functional roles
Personal authority	Positional authority	personal authority
Contract relations	Status relations	status relations

Each dichotomy can be represented as an edge or perpendicular bisector of a triangular opportunity space (Figure 4.2). For example, organic solidarity and specialized roles are positioned on the edge opposite the Egalitarian corner, which includes their dichotomous opposites, mechanical solidarity and multi-functional roles. This, in our view, represents the minimum framework or skeleton for a soft-bodied approach to environmental security. At a typical conference on development, advocates of all three approaches can be heard. 'Putting the poor first' encapsulates the idea of rational allocation of resources according to need (a hierarchical approach). 'Empowering communities for self-management' is an egalitarian point of view. These are countered by the reminder of an imperative to 'integrate into global markets'.

First, solidarity can be expressed through the *market*, characterized by the features of individualism and competition associated with *Gesellschaft*. Solidarity is achieved in two ways, most obviously through *contracts*, but also through individual consumption choices that establish identity with fellow consumers and differentiation from those who follow different consumption patterns. As manifestations of *Gesellschaft*, market forms of solidarity are directly orthogonal to both of the other basic modes of solidarity, described below, which share the stronger community boundaries typical of *Gemeinschaft* solidarity.

Second, solidarity can be expressed through orderly differentiation in *hierarchies*, which establish identity through careful gradations of *status* based on explicit characteristics such as age, gender, educational attainment, professorial rank, and so on. This form of *positional* authority is directly orthogonal to the emphasis on *personal* freedom shared by both markets and the third form of social solidarity.

Third, solidarity can be expressed through *egalitarian* homogeneity; that is, by operating rules of equality that keep each participant

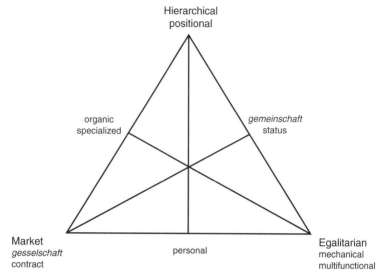

Figure 4.2 A two-dimensional space is required for complex strategy switching (Rayner, 1995)

at the same *status*. In this respect, egalitarianism is a manifestation of *mechanical* solidarity and *multifunctional* roles that is directly orthogonal to both hierarchies and markets which both favor *organic specialization* of labor.

Each model of social solidarity emphasizes different social arrangements for managing society and nature – including three different principles for trust and fairness. These social arrangements are briefly sketched out, for purposes of comparison, in Table 4.3.

As the above table indicates, each type is strongly associated with a different view of nature and therefore with different strategies for managing the environment. For example, the egalitarian view, that nature is fragile and can be destroyed by human actions, is expressed in the Precautionary Principle.

These diverse characteristics provide each strategy with a unique viewpoint that is both facilitating and limiting: *Hierarchies* are societal gardeners, experts in system maintenance so long as the garden is not disrupted by catastrophic disruption from outside. But, when unchecked by self-organizing groups and markets, hierarchies become corrupt. *Egalitarians* are societal canaries; they provide early warning systems of external dangers as well as of internal corrup-

Table 4.3 Three institutional strategies (Rayner, 1995)

	Market	Hierarchical	Egalitarian
Sovereignty	Consumer	Institutional	Group
Decision procedures	Skills	Rules	Consensus
View of nature	Benign	Perverse/tolerant	Ephemeral
Diagnosis of cause	Pricing	Population	Profligacy
Policy bias	Libertarian	Contractarian	Egalitarian
Policy instrument	Incentives	Regulation	Prohibition
Distribution	Priority	Proportionality	Parity
Trust	Successful individuals	Procedures	Participation
Liability	Loss spreading	Deep-pocket	Strict fault
Consent	Revealed	Hypothetical	Explicit
Timeframe	Short	Long	Compressed
Intergenerational responsibility	Present>future	Present=future	Future>present
Discounting	Diverse/high	Technical standard	Zero/negative

tion. They also tend to be repositories of local knowledge that is disvalued by hierarchies and overlooked by markets. But left to themselves they are prone to factional squabbling. Markets at their best provide the potential for innovation; they generate new ideas to resolve the problems created by the solutions of preceding generations. But, unchecked by hierarchy and self-organization, they are prone to extortion.

To function well, any of the three need the other two. Complex, overlapping, plural, interdependent civic institutions embodying diverse combinations of the three basic strategies represent the best means available to pressure and extend society's capabilities to develop in a sustainable fashion, even – or, rather, especially – when confronted with surprise. Strategy switching within these systems can be very rapid.

Conclusions

National and environmental security depends, not simply on hard-shell economic prosperity or technology, but on a society's soft-shell ability to switch strategies as global and local conditions change. Participation in multiple associations and the growth of trust in non-familial network relationships help build a secure society because they build the capability for complex strategy switching.

In pursuing any societal or environmental goal, we choose one of a complex set of strategies and social solidarities, thus emphasizing a certain viewpoint and precluding others. If we always choose the same strategy, we cannot build a secure society. But if, through a rich associational life and high levels of trust, we choose different strategies as conditions change, we are more likely to avoid destitution and work to eliminate poverty in our societies. Hence, the capacity for strategy switching is greatly enhanced by the coexistence and interaction of diverse forms of social solidarity (variegated social capital).

Hence, local development resources should be invested in creating and supporting a wide range of social activities that appear to have little relevance to providing basic needs, national infrastructure, or technical capacity building. Examples of such activities include football clubs, carnival societies, cultural associations, hobbyist clubs (gardening, photography, etc.), professional societies, craft guilds, scientific organizations and musical and theatrical groups.

Moreover, nation-building efforts should be redirected away from measures designed to homogenize populations and towards measures that actively encourage as much cross-cutting heterogeneity as possible.

At the global level, scientists should pursue their efforts towards building an international civil society through networking on a multitude of issues, interests, and identities (O'Riordan et al., 1998). This international society can be the source of complex viewpoints enabling strategy switching appropriate to national, international and global environmental security issues.

The capability for complex strategy switching grows out of multiple and overlapping associational ties within a society. Like an endoskeleton, a strong civil society may be more vulnerable to minor attacks than a society with heavily guarded borders. But it is also more capable of responsive adaptation to major assaults in a world characterized by rapid social change. It is therefore more resilient and more sustainable, better able to ensure its environmental security.

Notes

1 However, care is required in defining stability. Post-war Italy provides a vivid case in point. Measured by formal political science indices, such as the number of elections or turnover of prime ministers, Italy has been judged to be politically unstable. However, from the point of view of continuity in power of the dominant party and personalities, this

seems to have been merely a game of musical chairs. The dominant political power relationships in post-war Italy have remained intact and unchanged for half a century.
2 Interestingly, the conventional use of the term 'social security' refers to securing people's economic welfare through insurance mechanisms. There is precious little that is 'social' about it, and it continues to rely on the assumption of economic determinism.

5
Human Security, Environmental Security and Sustainable Development

Steve Lonergan

> Poverty is a major cause and effect of global environmental problems. It is therefore futile to attempt to deal with environmental problems without a broader perspective that encompasses the factors underlying world poverty and international equality.
>
> WCED, 1987

> The environmental problems of the poor will affect the rich as well, in the not too distant future, transmitted through political instability and turmoil.
>
> Gro Harlem Brundtland, 1986

Introduction

Although it has become commonplace to quote from the World Commission on Environment and Development's report 'Our Common Future', the two statements above (the second by the Chair of the Commission) highlight three of the key issues facing the global community as we prepare to enter the twenty-first century: environmental degradation; impoverishment; and the insecurities caused by both of these.[1] Over the past decade, there has been increasing discussion of the links among these three issues, and between environment and security in particular. Many of the authors who have been central to this discussion are represented in this volume. Accompanying the environment and security debate has been a lively dialogue on the nature of security and whether the 'threats' to security are now different than they once were. This discussion has focused on the declining relevance of traditional military conceptions of security and the need to 'redefine security' to include non-conventional threats. Non-conventional threats include religious fundamentalism, resource scarcity, human rights abuses, outbreaks of infectious disease, and environmental degradation caused by toxic contamination, ozone depletion, global warming, water pollution, soil degradation and the like. Much of the early literature on environment and security was general and anecdotal (cf. Brown, 1977; Myers, 1986, 1989; Mathews, 1989; Renner, 1989). Following this, a number of authors have tried to develop more rigorous arguments linking environment and security by concentrating specifically on the role of environment and resource depletion as potential *causes of violent conflict* (Homer-Dixon, 1991, 1994; Libiszewski, 1992; Bächler, 1994). Such conflict, in turn, could pose a serious threat to the security of individuals, communities, states and regions. The general discussions on the nature of security and the role of environmental degradation as a contributor to insecurity and conflict have been labeled by Levy (1995a) as the 'first wave' of environment and conflict research. Subsequent research attempting to 'prove' a link between environment and conflict was considered by Levy (1995b) to be the 'second wave'.

Discussion of the links between environment and security has not been confined to academic circles; Warren Christopher, former US Secretary of State, has explicitly linked the two, noting that '... natural resource issues (are) frequently critical to achieving political and economic stability'. Whether the debates have been among academics or in public policy circles, there is little doubt that a considerable amount of confusion has been generated over how environment and security are 'linked'. As Dabelko and Simmons (1997) note, the diversity of conceptual perspectives persists not only within disciplines, but within government departments as well. Indeed, many researchers avoid using the term 'security' altogether, preferring to focus on environmental change and social adaptation and/or armed conflict.

The resulting muddled discussion can be partially attributed to the various interpretations of 'environmental security'.[2] The interpretations can be divided into four categories, as follows:

1 Security of the environment (or security of services provided by the natural environment; this has also been interpreted as non-diminishing natural capital)

2 Environmental degradation and resource depletion as potential causes of violent conflict;
3 Environmental degradation and resource depletion as threats to national welfare (and, therefore, national security); and
4 Environmental degradation and resource depletion as two of many integrated factors which affect 'human' security.

These discussions have been more obscure due to their interdisciplinary nature; participants range from environmentalists, who interpret a 'secure environment' as no further decline in the stocks of natural capital, to military specialists, who view environmental security as a 'greening' of their operations. Adding to the confusion are researchers who are largely ignorant of the traditional security debates and who often interpret 'security' very differently from members of the international relations community or the defense establishment. Conversely, the introduction of new environmental terms by the political science community (such as 'environmental scarcity' or 'environmental refugees') only frustrates researchers and policy makers for whom similar terms have specific, important and sometimes legal meanings.

To be fair, the effort to link environment and security has not been unconditionally accepted by all researchers and policy makers. Deudney (1991) argues that promoting environmental degradation as a threat to security is largely rhetorical – an 'old horse' (security) attached to a 'new wagon' (the environment). However, the muddled debate continues when Deudney and others warn against the 'securitization' of the environment rather than the greening of security if the 'mismatched tools' of military institutions are to play a role in 'securing' the environment or countering environmental threats (Dalby, 1992; Conca, 1994; Käkönen, 1994; Wæver, 1996).

The purpose of this chapter is to present a new and different perspective on the environment and security link from those which have been the purview of the political science and international relations communities. It argues for a more 'interdisciplinary' and 'integrative' perspective on the issue, by focusing on three key premises. First, human perceptions of the natural environment, and the way we use the environment, are socially, economically and politically constructed; any attempt to remove these other factors from the discussion and consider the environment as a causal factor in conflict simply ignores reality. Second, we must recognize that environmental problems must always be addressed from a broader perspective that encompasses both world poverty and issues of equity (intergenerational, international and intra-national; see Lonergan, 1993). This was a clear recommendation of the World Commission on Environment and Development (1987), and implies that when talking about resources and the environment, one cannot ignore issues of equity and impoverishment. And third, we must accept that 'space matters'. That is, the appropriate spatial level in which to deal with both environmental and security concerns is not necessarily the nation-state, but must be directed towards where the knowledge base is the greatest (often the local level).

These three premises greatly influence the discussion below. And while I believe that linking environment and security is ultimately a useful endeavor, I also believe that less attention should be given to researching the links between environment and *violent conflict*, and more be devoted to obtaining a better understanding of how environmental change is related to *human security*.

Defining 'human security'

Initially, human security was interpreted as meaning threats to the physical security of the person. For example, the Universal Declaration of Human Rights, adopted by the UN in 1948, states that 'everyone has the right to life, liberty and the security of person . . .'. However, the concept now encompasses economic, health, and environmental concerns as well. It is, as the UNDP (1994) notes, an 'integrative' as opposed to merely a 'defensive' concept. Their definition of human security includes seven categories of threats:
- Economic security (assured basic income)
- Food security (physical and economic access to food)
- Health security
- Environmental security (access to sanitary water supply, clean air and a non-degraded land system)
- Personal security (security from physical violence and threats)
- Community security (security from ethnic cleansing)
- Political security (protection of basic human rights and freedoms)

This concept of human security has a spatial component as well. The UNDP recognizes global challenges to human security, which arise because the threats are international in nature. Included in threats to global human security are:
- Unchecked population growth
- Disparities in economic opportunities

- Excessive international migration
- Environmental degradation
- Drug protection and trafficking
- International terrorism

The appeal of the term 'human security' is that it recognizes the interlinkages of environment and society, and acknowledges that our perceptions of the environment, and the way we use the environment, are historically, socially and politically constructed.

Environment, security and sustainable development

Reports from the World Commission on Environment and Development and the United Nations Conference on Environment and Development held in Rio de Janeiro in 1992, along with proceedings from various other meetings, have all reinforced the notion that economic development is absolutely necessary to improve quality of life and well-being. However, to be sustainable, this development must incorporate environmental, social, political and economic considerations consistent with a strong commitment to equity and a concern for resource limitations and the viability of ecosystems. Indeed, the social, political, economic and environmental are inter-linked; any actions involving one of these systems necessarily affects the others. Even our *perspective* on the environment is constrained by other social, economic, political and cultural factors. Why then should we be adopting a model from the sciences to study environment and security or conflict – which promotes linear cause and effect relationships – to determine causality when this is flawed to begin with? The past decade of discussions on sustainability have taught us that any assessment of public programs and politics must take into account how outcomes of government interventions impact upon sustainability and how such outcomes may promote or upset the balance between development and the environment.

A particularly relevant criticism of the existing literature is that the debate on how environment and security are linked has largely ignored a discussion of how environmental security fits into the broader context of international development, social welfare and sustainable development. While there have been calls for attention to 'comprehensive security' (Westing, 1986) and a 'multilevel approach' to security (Græger, 1995), most authors fail to recognize the one overwhelming argument in favor of linking environment and security: *that environmental problems must always be presented from within a broader perspective that encompasses various forms of inequity, including world poverty.* Why is this the case? It is the case precisely because poverty and inequity are two of the key factors contributing to tension and insecurity throughout the world. And while this premise is related to the one above, it also stands as more of a challenge to those working on environment and security issues. The importance of presenting environment and security issues in the context of inequity is highlighted in the last section of this chapter which focuses on international tensions over access to fresh water. Most discussions that examine conflicts over water argue that disputes either result in conflict or cooperation; often relevant discussions over inequities in accessing resources are ignored.

The third premise is that 'space matters'. Some authors have recently called for a reshaping of the research on environmental security which, in turn, would affect how we would define and apply the concept. Two particular trends are worth noting. First, there has been a strong plea for *regionalism* (sub-national and supra-national) which should be decided on the basis of eco-geography (Dokken and Græger, 1995). Here eco-geography implies that the region of concern is defined by ecological boundaries (eg watershed or eco-region) and not simply political boundaries. I would argue that regionalism need not be confined to ecological regions; it may incorporate ethnic or other factors as well. Perelet (1994) also argues that environmental security problems must focus on the ecosystem level. Second, there has been a need expressed to focus research at the local level, or at least at the lowest possible spatial level for the insecurity/conflict being studied or the action being proposed. This has been termed the 'subsidiarity principle' (Mische, 1989; Dokken and Græger, 1995). For example, the relationship between environment and poverty must be viewed locally and in the context of broader human security dimensions. This is absolutely crucial if we desire to incorporate local knowledge and make the process of development a truly participatory one. On the other hand, some actions are most appropriate at the global level (eg global warming agreements). These two trends have stimulated further questions on the relationship between environmental security and the goal of a sustainable society.[3]

Environmental change and human security

So where does this leave the link between environment and security? Coincidental with the claim that environmental degradation may be a threat to security has been the move to interpret security more broadly to encompass notions of human welfare, where there is no immediate prospect of armed conflict. A better way of framing the issue might be in terms of *insecurities* (Græger, 1996; Lonergan, 1997). As noted above, the appeal of a broader notion of security, framed in terms of insecurities, is obvious. It allows one to include the key issues of equity and impoverishment, along with other non-conventional threats to security. It also is multi-dimensional (or 'multi-level' according to Græger, 1996); this allows us to introduce into the debate issues of rights and responsibilities, acting at all spatial levels, from the individual to the global. However, as Soroos (1994) notes, many authors have expressed significant concerns with using a broader definition of the term 'security'. These concerns include:

- the concept of security loses clarity and meaning when it is used broadly to include threats other than those of a military nature;
- environmental insecurities bear little resemblance to military threats and thus are dealt with in fundamentally different ways (the 'mismatched tools' argument of Deudney, 1991; and Wæver, 1996);
- the traditional conception of security as a human value is biased toward preservation of the status quo, while revolutionary social changes are needed to address environmental change;
- measures taken in the name of environment and security will perpetuate economic and social injustices, because if viewed as security threats, environmental threats may reinforce nationalistic sentiments and the state system, and may become an excuse for undemocratic tendencies.

Even accounting for these concerns, it is clear that there is a need to reassess our traditional perspectives of security and, I would argue, focus more on the 'insecurities' posed by non-conventional threats. We must also recognize that society is in a period of rapid social, economic and environmental change. While continued growth in gross world product may imply improvement in societal welfare, there are many disturbing indications that this is simply not the case for much of the world's population. The number of wars has increased dramatically over the past three decades. However, since 1960, these have primarily been intra-state conflicts, where inequity is a key feature. The spatial element has become a crucial factor in this changing nature of conflict. We now face the specter of global warming. The income gap between rich and poor is widening. Population continues to increase and has become more urbanized. And terrorism, pollution and disease are all on the rise in many regions of the world. It has also been suggested that there has been a decline in the level of trust within and between regions, and that this may affect not only economic output, but has also been a contributor to the decline in social capital[4] in most regions of the world (Fukuyama, 1995). These actions reinforce the broader sense of an inequitable society that is at the ultimate root of the insecurities experienced by individuals, communities and nations.

This leads to the inescapable conclusion that while environmental degradation and resource depletion as causes of conflict may be overstated, it is undeniable that increasing inequities in society are likely the major source of these (and other) non-conventional threats to individual security. At the same time, there is recognition that traditional approaches to national security may not, in turn, ensure the security of individuals and communities. This applies to global efforts aimed at curbing global warming and ozone depletion as well as to local anti-crime initiatives. Again, this calls for a broader notion of security, one that focuses on the 'human' element, despite the ambiguities and difficulties this may cause.

Indeed, such ambiguity is more common than most realize. Almost no term used to describe major sectors of human activity is without ambiguity or does not overlap with many other issues. While some may feel a term like 'sustainable development' is merely '... a catchword, bandied about by all kinds of lobbies to achieve all sorts of goals' (Græger, 1995), the fact remains that it has galvanized disparate groups into working together for a common goal: ensuring that our social, economic and ecological systems are sustainable.[5] In spite of debates about the meaning and use of the term, discussions of 'sustainable development' have resulted in an enormous amount of research and policy design which recognizes the links among environment, economy and society. Such debates have also enriched the research area, and have resulted in at least a perception of sustainable development that is inclusive of a variety of perspectives. As Soroos (1994) notes, many frequently used concepts – such as peace, conflict and justice – are intrinsically abstract and inevitably subject to a variety of interpretations. However, the usefulness of these concepts is not in conveying a precise meaning, which would render them intellectually barren, but in the discussions and controversies they provoke, which lead to new insights and perspectives. Environment and security, particularly when security is broadly interpreted to imply human security, may have a similar effect. In addition, the differences between military and environmental threats do not imply that linking environment and security is a meaningless exercise; rather, it suggests that an examination of the similarities and differences between these different types of threats could result in a more complete understanding of both.

Human security and sustainable development

Adopting a broader notion of security which encompasses issues of equity and impoverishment – what has been termed 'human security' (see Figure 5.1, above), raises the question whether focusing on environmental change and human security does little more than propose the adoption of a term to replace sustainable development. There is little question that human security and sustainable development are related; indeed, the former might be considered a precondition for the latter.

As defined by the World Commission on Environment and Development, sustainable development is '... development which meets the needs of the present generation without compromising the ability of future generations to meet their own needs (WCED, 1987)'. This concept has now been embraced by individuals, communities, governments and institutions and, similar to the notion of environmental security, research and writing ranges from the most general and anecdotal to specific cases of sustainable communities. One main distinction between human security and sustainable development is that the former is primarily *instrumental or analytical*,[6] while the latter is primarily *normative*. That is, with the exception of the general discussion (Levy's 'first wave'), the focus in environment and security research (however defined) has been on root causes, while much of the emphasis in sustainable development has been on developing guiding principles with which to inform policy.

A slightly different way of saying this is that human security focuses on individual or collective human perceptions and evaluations of actual and expected conditions of the environment as a source of insecurity, while sustainability focuses on changing the way we think about the environment. In this sense, the framework implied by a human security perspective is a *participatory* one; those being researched become the researchers, and vice-versa.

The early literature on sustainable development explicitly acknowledged the role of poverty as a major cause and effect of global environmental problems. This recognition is also explicit in the environmental security literature. However, substantive policy responses (or even discussions) have not accompanied these assertions on the importance of including impoverishment and inequality within the context of sustainable development. While much writing on sustainable development recognizes the important links between environment, economy and society, the main focus has been on the environment – economy tradeoffs. 'Sustainability' has very little meaning in the context of cultural diversity or violence against women. On the other hand, environment and human security recognizes the integrated nature of various aspects of the human existence and the natural environment.

More importantly, however, the notion of human security encompasses the dialectic of *rights and responsibilities*. This dialectic is applicable at all spatial levels, from the individual to the global community. These rights and responsibilities need to be recognized not solely in the context of whether future generations will be able to achieve the same level of development as the present generation, but *within* the existing generation as well. The guidelines which outline our rights and responsibilities have been labeled by some as the 'terms of agreement'[7] which affect the relationship among individuals, cultures and nations. What are 'terms of agreement?' In the environmental arena, one immediately thinks of agreements such as the Climate Change Convention or the Biodiversity Convention, both signed at the UN Conference on Environment and Development in 1995. A critical analysis of such agreements may uncover the deficiencies in both the agreements themselves and the processes leading to the signing of these agreements (Narain 1997), particularly with respect to the rights and responsibilities embodied in each. Are there similar 'terms of agreement' with respect to human security? In my mind, the overarching terms were set out in the Declaration of Human Rights adopted by the United Nations in 1948, which stated: '... everyone has the right to life, liberty and the *security of person*...' (my italics). It is clear from this document that human security is – and should be – a cornerstone for linking issues of environment, economy and development.

Adopting this dialectic of rights and responsibilities as a framework implies that we are no longer stuck with the deterministic conceptions of uni-causality that has plagued much of the work on

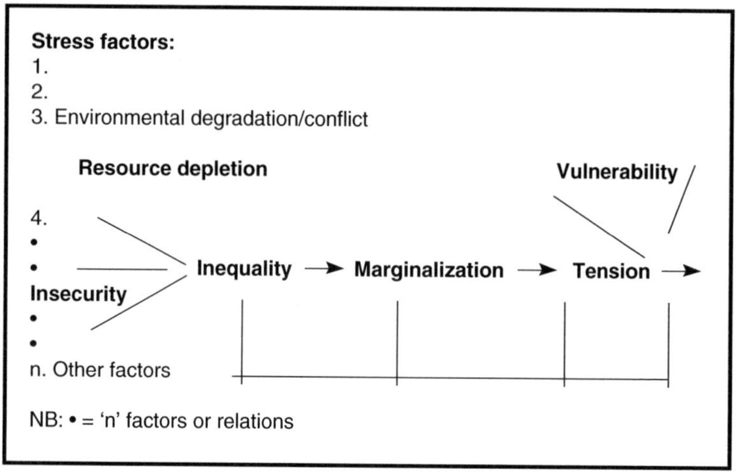

```
Stress factors:
1.
2.
3. Environmental degradation/conflict

         Resource depletion                    Vulnerability
4.
•
•  ─────────  Inequality  ──►  Marginalization  ──►  Tension  ──►
Insecurity
•
•
n. Other factors

NB: • = 'n' factors or relations
```

Figure 5.1 Conceptual diagram of environment and human security linkages.

environment and conflict. How, then, can one conceptualize the process of thinking and researching within a framework of environment and human security? The following conceptual diagram (Figure 5.1) may be helpful in this regard.

Three issues are implicit in the diagram. First, the relationship between the environment and human security is a complex one. The role of environmental change and resource depletion as a contributor to insecurity is affected by many other factors (termed 'stress factors' in the diagram). In many cases, environment may play a minimal role in contributing to insecurity. It is also likely that there exists a cumulative causality, with significant feedback mechanisms. Therefore, it is impossible to remove one factor from the overall historical, spatial, cultural, political context.[8] Indeed, we may gain a better understanding of the process by examining cases where environmental degradation occurred, but insecurity did not result. Second, the implications of each component in the diagram range across a spectrum of social, political and environmental activities. That is, both cause and effect are multi-dimensional. Last, although depicted in a direct, linear fashion, the links among the various components may be direct or indirect.

In the diagram, inequity is depicted as the root cause of insecurity, often appearing as impoverishment, and tempered through the stages of marginalization (which includes dual components of disenfranchisement and disengagement, or breaking off communication) and increasing tension. Various external stresses or threats influence the process, as does the degree of vulnerability (defined as a measure of coping abilities). All of these components must be interpreted within the context of prevailing terms of agreement, and the rights and responsibilities embodied in these terms.

The example below illustrates the linkages outlined in Figure 5.1 in the context of water in the Jordan River valley. While much has been made about the 'conflict' over water throughout the region, resource scarcity *per se* is not the cause of insecurity. Inequitable access to resources, discriminatory pricing policies, and lack of control have been much more important. As Israeli hydrology professor Uri Shamir noted, 'if there is a political will for peace, water will not be a hindrance. If you want reasons to fight, water will give you ample opportunities' (as cited in Vesilind, 1993). The insecurities that surround the access to water by the Palestinians are the result of inequitable actions and policies on the part of Israel; inequities that extend beyond simple access to water, and include access to capital and other resources as well (Lonergan and Brooks, 1994).

Water in the Israeli-Palestinian context[9]

The Jordan River basin presents an interesting case for examining the relationship between environment and human security, as the region exhibits all of the components identified in Figure 5.2, above. Central to the tensions that exist between Israel and the Palestinians is the availability of adequate fresh water supplies. In addition to the obvious water scarcity/conflicts problem, the existence of refugees – Palestinian, Ethiopian, Russian and others – stresses political, social and environmental systems. There are also significant constraints on the level of economic achievement of certain sectors of national or regional economies due to a lack of resources, the mining and deterioration of the groundwater supply, and the control of resources by a number of states in the region. The situation has become so extreme that King Hussein of Jordan noted water as the *only* issue that would lead him to go to war with Israel. Despite the recent advances made in the peace discussions, the water issue remains a major stumbling block to a lasting peace in the region.

Inequities

Virtually all of Israel's fresh water comes from two sources: surface water supplied by the Jordan River, or groundwater fed by recharge from the West Bank to one of three major aquifers. There is a long legacy of controversy over fresh water in the region, dating back thousands of years. In recent times, there was a proposed comprehensive plan for cooperative use of the Jordan River (the Johnston Plan) as early as the 1950s, but this was derailed by mistrust among the four riparian states (Israel, Jordan, Lebanon and Syria). Each nation has tended to follow its own water policies since the failure of that agreement, often to the detriment of other nations.

Water has long been considered a security issue in the region, and on numerous occasions Israel and its neighboring Arab states have feuded over access to Jordan River waters. Many authors have argued that a major contributing factor in the tensions leading to the 1967 war was the water issue. At the time, Israel was consuming almost 100 per cent of its available fresh water supplies. Occupation of the three territories (the West Bank, the Golan Heights and the Gaza Strip) after the war changed this situation in two ways. First, it increased the fresh water available to Israel by almost 50 per cent. Second, it gave the country almost total control over the headwaters of the Jordan River and its tributaries, as well as control over the major recharge region for its underground aquifers. Control of water resources in the West Bank and the Golan Heights are now integrated into Israel's economy and, accordingly, essential to its future.

Presently, Israel draws over 40 per cent of its fresh water supplies from the West Bank alone, and the country would face immediate water shortages and a significant curtailment of its agricultural and industrial development if it lost control of these supplies. Former Israeli agricultural minister Rafael Eitan stated in November of 1990 that Israel must never relinquish the West Bank because a loss of its water supplies would 'threaten the Jewish state'. The growing number of settlements in the region poses an additional problem. The water in the West Bank is now used in a ratio of 4.5 per cent by Palestinians and 95.5 per cent by Israelis (while the population is over 90 per cent Palestinian). The UN Committee on Palestinian Rights concluded in 1980 that Israel had given priority to its own water needs at the expense of the Palestinian people.

To ensure security of water supply from the West Bank aquifers, Israel has put in place quite restrictive policies regarding Palestinian use of water. Israel's application of restrictions on Palestinian development and use of water not only improves its access to West Bank water, but also extends its control throughout the territory. It is this inequitable situation with respect to water allocations which increases resentment and adds to tensions in the region.

Marginalization

After the 1967 war, responsibility for water resources and water management was taken over by the Israeli military. Initially, the Civil Administration delegated its powers to a 'water office', which was, in turn, responsible to the Israeli Water Commissioner. In 1982, responsibility was transferred to Mekorot, the water supply agency, which collaborated only minimally with the Palestinians. All water activities are also subject to military orders, which have been extensive. Until very recently, all publication of information on water was prohibited. Palestinians have no access to water data.[10]

Tension

There is little doubt that the inequitable allocation and pricing of water has increased tensions between the Israelis and the Palestinians. Although the claim that water was a major cause of the 1967 war is much disputed (cf Naff and Matson, 1984; Lonergan and Brooks, 1994), there is little doubt that the development of Israel's National Water Carrier in 1964 and subsequent Syrian attempts to divert the headwaters of the Jordan River played a part in the chain of events leading to the war (Heller and Nusseibah, 1991; Lowi 1993, 1995).

Much of the tension over water between the Palestinians and the Israelis relates to the blatant discrimination in water pricing, allocation and delivery systems. Water consumption by Israeli settlers in the West Bank is roughly eight to ten times that of the Palestinians. Water is sold to Israeli settlements for 0.5 New Israeli Sheckels (NIS) per cu.m. (settlers in Gaza pay only 0.3 NIS), while it is sold to Palestinian villages for 1.8 NIS per cubic meter (and 2.1 NIS to individual Palestinians). Water is available to Palestinian villagers only one or two days a week (and is otherwise stored in water tanks on the roofs of houses) while it is made available daily and on demand to Israeli settlements. These discriminatory practices are enforced through the application of Israeli military orders to the West Bank and Gaza. Coupled with the lack of agreement on water rights – or the Israeli unwillingness to discuss issues of rights of control – these inequities have increased tensions throughout the region.

Insecurity

Arabs in the West Bank have protested for years to Israeli authorities that their agriculture and economy are being negatively impacted or even ruined by unfair water policies, and that the water network supplying the Jewish settlements has drastically depleted the villages' water resources. However, there has been no outlet for the Palestinian complaints. While some of the Palestinian charges are disputed, the issue is a highly emotional one, and a number of authors have used the metaphor that the water situation in the Occupied Palestinian Territories is a 'time bomb waiting to explode'.

Terms of Agreement

On September 28, 1995, Israel and the Palestinians signed the Taba Interim Agreement ('Oslo II Accords') as part of the peace process which began in Madrid in 1991. The Interim agreement[11] contained provisions for future relations between Israel and the Palestinians, and called for the establishment of a Palestinian Council and permanent status negotiations (which would begin by May, 1996). One of the key provisions (Article 40) of the agreement concerned water and sewage. Some of the key elements relating to this Article include the following:

1. Recognition of Palestinian water rights in the West Bank. However, these 'rights' were not defined and will be negotiated in the permanent status talks. What is clear is that Israel interprets these as rights to *amounts* of water, while the Palestinians interpret these as rights to *control* over water.
2. The parties agreed to cooperate on the joint management of water and sewage systems in the West Bank.
3. Co-operation in water-related technology transfer, research and development, and the development of future sources of supply.
4. Acceptances of the future water needs of Palestinians in the West Bank of double the present allocation.

Although the Taba Agreement is only an interim agreement (of 400 pages), the terms set out in the Agreement – particularly with respect to the issue of water rights – remain unfavourable to the Palestinians. In addition, the election of Benjamin Netanyahu as Prime Minister of Israel in the spring of 1996 raises the question of whether there will every be a permanent agreement signed between the parties which respects the rights of Palestinians. The Likud-led conservative coalition in Israel has consistently ignored the spirit – if not the letter – of the Oslo Accords, with continued development of settlements in the West Bank and intermittent closures of the Territories, preventing Palestinians from entering Jerusalem or working in Israel. With the establishment of the Palestinian National Authority, there continues to be an expectation that the Palestinians will be allowed to run many of their own affairs on the West Bank. However, it has been made clear in recent pronouncements by the Prime Minister and members of his cabinet that foreign affairs and security, and *water resources* will continue to be controlled by Israel.

Conclusion

What is the most appropriate conception of security? What spatial scale does this conception operate at? And how does environment and human security relate to sustainable development? This chapter has attempted to answer these questions and present a framework for the future assessment of environment/security linkages. The answer to the first question is that any acceptable notion of security must begin with an understanding of equity. This implies that when assessing the implications of environmental degradation and resource depletion on security, the appropriate framework should be one of 'human security'. While not a new concept, its importance is magnified by the scale of environmental degradation and other non-conventional threats to security, the increasing disparity between the rich and the poor, and the realization that ensuring national security no longer ensures the security of individuals and communities.

The second question – one of appropriate spatial scale – is easily answered within the context of human security. Since national security no longer ensures the security of individuals and communities, in the context of human security the most appropriate spatial scale is the 'smallest' one. That is, there must be a decentralized approach to resolving the problems of non-conventional threats to security, whether environmental or otherwise. Hence, the principle of subsidiarity mentioned above is implied. This also mandates a participatory approach to examining environment/security linkages. One important lesson from the institutional processes created to focus on sustainable development issues is that local level knowledge is extremely important. Whether we term it 'traditional ecological knowledge' (TEK), or simply local knowledge, understanding how environment interacts with social, economic and political systems must be understood at all levels, from the local to the global.

Last, it is clear there are significant differences between sustainable development and the above model of environment and human security. In particular, human security focuses on individual or collective human perceptions and evaluations of actual and expected conditions of the environment as sources of insecurity, while sustainability focuses on changing the way we think about the environment.

I have also tried to emphasize that adopting a human security approach to assessing environment/security linkages embodies the important dialectic between rights and responsibilities. This dialectic should be used to assess the social relevance of any terms of agreement and to guide research and policy development in the future. It also becomes clear that a conceptual framework which stresses environment and human security bridges the gap between the study of environmental change in the context of traditional conceptions of security and the notion of sustainable development.

Notes

1 I would like to acknowledge and thank Pamela Moss, Geoff Dabelko, Denise Pritchard and Andrea Blower for their comments and suggestions on earlier drafts of this chapter. I would also like to acknowledge Sunita Narain of Delhi, India, whose remarks helped shape the ideas expressed herein.
2 Although some of the literature draws a distinction between 'environment and security' and 'environmental security', in this chapter they are used synonymously.
3 It should also be noted that there are equity issues which may be place specific, but may be lost in adopting a simple spatial perspective – for example, class, gender and citizenship. The claim that 'space matters' is not an attempt to ignore these issues, but to highlight the importance of the spatial dimension.
4 Social capital is the ability of individuals to organize for a common good.
5 It appears this may be an exception to the general decline in social capital noted above.
6 Here I use 'instrumental' in the sense of Habermas' instrumental versus symbolic values. That is, instrumental implies the search for a better specification of the links between environment and security, while normative approaches focus on policy prescriptions.
7 These thoughts on terms of agreement and the dialectic of rights and

responsibilities are based on discussions I had with Sunita Narain of Delhi, India. These ideas were formulated in conversations leading up to our presentations at the Second Open Meeting of the Human Dimensions of Global Change Community, Laxenburg, Austria, June, 1997.
8 Unquestionably, this poses a dilemma for researchers, since one is always limited in what one sees, or includes. However, it is important to recognize that each factor is 'embedded' in other sets of structures.
9 For more information see Naff and Matson, 1984; Lowi, 1995; Lonergan and Brooks, 1994.
10 This situation was partially rectified in the 1993 Interim Agreement.
11 The 1993 Interim Agreement set the stage for final status negotiations. At this time, tensions over Israeli settlements in the West Bank and recent terrorist attacks against Israel have resulted in a suspension of all discussions between the Palestinian National Authority and the Israeli government. The prospects for peace at this point in time seem dim, indeed.

6
Geopolitics and Ecology: Rethinking the Contexts of Environmental Security
Simon Dalby

Environmental security

The end of the cold war has caused considerable policy and scholarly attention to be devoted to sources of political conflict other than those generated by major power rivalry.[1] Concerns with global climate change, ozone depletion and acid rain have sometimes made matters of global environmental security an international political priority (Soroos, 1997). There is also a growing scholarly and policy literature on relationships between environment, renewable resource scarcities and violence, driven in part by fears of political instability and state collapse as a result of environmental degradation and resource scarcity (Gleditsch, 1997a; Diehl, 1998; Renner, 1996).

These concerns are not necessarily especially new, but in the 1990s they have been getting attention both as part of the process of international diplomacy concerned to regulate climate change and other environmental matters (Elliot, 1998; Paterson, 1996), and as part of the emergence of a trans-national discourse of 'environmental security' (Deudney and Matthew, 1999). Although much of the literature concurs that there is little immediate likelihood of environmental degradation causing inter-state warfare, concerns over environmentally induced political instability and politically disruptive migration continue to engage foreign policy organizations in Washington.

Clearly the international security agenda has changed through the last few decades, but, despite considerable discussion, exactly how this new found concern with environment should be related to security is still much less than clear. Both the research work on the 'environmental degradation leading to conflict' hypothesis, much of which has focused at the intra-state level (Homer-Dixon, 1994), and the larger policy frameworks of environmental security, have come in for considerable criticism. The critiques challenge the conceptual frameworks used in both the scholarly research and policy discussion (Brock, 1997; Deudney, 1997; Gleditsch, 1998; Levy, 1995). These debates raise important doubts as to whether the current formulations of environment in international relations thinking, which, among other things, often consider environmental security as a matter of state responses to global change, are conceptually adequate.

The debate about definitions of security, and the identification of environmental sources of political threat, is part of the larger post-cold war scholarly discussion of the appropriate concepts and methods for the study of international security (Krause and Williams, 1997; Buzan, Wæver and de Wilde, 1998). Drawing on these theoretical discussions within the international relations literature, the argument that follows here raises two substantive concerns about environmental factors as a matter of international security: First, the post-cold-war claims that environment is a new matter for security analysis raises the question of the appropriate historical context in which to think about matters of environment and conflict; Second, given current criticisms of the geographical and spatial premises of international relations thinking (Agnew, 1998; Ó Tuathail, 1996; Walker, 1993), just what are the appropriate geographical frameworks for investigating matters of environmental change and conflict?

This chapter briefly investigates these two interrelated themes to show that adding the historical dimension to the discussion, and thinking it through in the largest of geopolitical frameworks, adds an important critical edge to the understanding of the current debate.

Environmental geopolitics

The most high profile articulation of the themes of environmental security is probably Robert Kaplan's 'coming anarchy' article published as the cover story to the February 1994 issue of the *Atlantic Monthly*. Kaplan raised the specter of political chaos induced by a Malthusian mismatch between population growth and resource inadequacies. Using the political failures of West Africa as his suggested model of the future, this argument constructed the world in imaginary spaces of chaos and anarchy contrasted to the affluence of the developed world (Dalby, 1996). The danger lies, he argued, in the spillover of chaos from the zones of turmoil into the zones of peace. The most obvious dangers are in the form of drug trading, disease and large-scale transnational crime which will merge with new forms of warfare and terrorism. This division of the world into separate geopolitical regions is important to both Kaplan's argument and much of the rest of the literature on environmental security where the underdeveloped global 'South' is portrayed as the source of instabilities that threaten the 'North' (Dalby, 1999).

But this conventional portrayal of geopolitical dangers obscures many of the important processes in the contemporary global political economy. It often obscures the simple point that environmental degradation is often geographically distanced from the elites who benefit from the commodities produced in the process of degrading environments (Gadgil and Guha, 1995). As the tragic case of the destruction of Ogoniland in Nigeria by oil exploration and extraction processes has recently been made especially clear, the destruction of environments related to oil wells and other production activities often occurs far from the places where the petroleum is finally consumed (Sachs, 1995; Watts, 1998). Other mining operations, and the disposal of at least some of Northern toxic wastes in states in the South, shows that the geography of pollution and degradation is much more complex than suggested in the simple models of zones of chaos and zones of peace.

The focus on geopolitical dangers rarely examines the specific causes of degradation in any detail. Often, in parallel with Kaplan, the environment is reduced to an exogenous factor or an independent causal variable in analyses of conflict. This approach is usually in danger of replicating the faults of environmental determinism in attributing simple causal power to natural environments in human affairs while disregarding the specificities of human institutions in particular places (Lipschutz, 1997). It also frequently overlooks the important cross-boundary patterns in the flows of resources and the politics involved in the dispossession of peoples to facilitate resource extraction and the expansion of commercial agriculture. (Dalby, 1999)

Many post-colonial societies are tied into patterns of export-led development that persist from colonial times. Forests are stripped in the South to make commodities available in the North. The

Japanese forestry example is frequently quoted; forests on the Japanese islands are relatively well cared for, but the supplies of timber for the Japanese economy come from South East Asia (Dauvergne, 1997). The resulting deforestation there has seriously damaged ecosystems across the region, destroyed the ecological base of survival for some indigenous peoples, and led to the environmental disasters of massive fires and air pollution in the late 1990s.

Thus, international resource flows are an important part of what needs to be understood in discussing the political ecology of global change (Bryant and Bailey, 1997; Redclift, 1996). Resource flows are obviously not the whole story, but often they do not appear in discussions of environmental security. Where they do appear in the international relations literature, often in discussions of trading regulations and regime formation, their significance for the large scale changes that are crucial to the politics of global environmental change are also obscured.

The 'territorial trap' in international relations thinking is clearly working here, at the largest scale, in specifying the environmental factors as within individual states. In John Agnew's (1998) terms, the territorial trap occurs as a result of three assumptions prevalent in international relations thinking. First, states have exclusive powers over their territories in terms of sovereignty. Second, domestic and foreign realms of state activity are essentially separate spheres of activity. Third, the boundaries of states define the boundaries of societies contained within those states. Combined, these three elements lead to a 'state-territorial' understanding of the workings of power.

Using state-territorial entities as the basis for analysis of the causes of environmental destruction focuses attention within the containers and in doing so offers a very selective geographical understanding of the processes of global change, resource use and degradation. The geography of population displacement and environmental degradation is much more complex than formulations of politics operating within the limitations of the territorial trap usually allow.

In his suggestions for improving the conceptual framework used to examine the relationships between environment and war, Gleditsch (1998) points to the necessity of connecting inter- and intra-state violence. But the obvious connection of non-violent action in one state having potentially violent consequences in others also needs attention. Whatever theories of environmental causes of violence are formulated, they need to incorporate this more complex understanding of cross boundary linkages to adequately grapple with the complexities of contemporary processes (Dalby, 1999).

Environmentally benign democratic peace?

While much of the literature on environmental security is pessimistic, if not downright alarmist at least in its popular articulations, some other contemporary international relations literature on warfare is much more optimistic about the future. Contemporary discussions of 'the democratic peace' suggest that advanced industrial democracies do not fight each other and that major wars are obsolete in the nuclear age (Russett, 1993). Coupled to the optimism that the end of the cold war will lead to the expansion of the number of democratic states in the world, this suggests that the future may avoid the major violent inter-state confrontations of the past.

When the 'democratic peace' argument is linked to considerations of the supposedly superior records of industrial democracies in environmental protection, it can be suggested that there is considerable hope for dealing with environmental security questions (Gleditsch, 1997; Midlarsky, 1998). The suggestion is that environmental problems occur more frequently in parts of the world that have authoritarian or communist regimes. Democratization should lead to environmental improvement by enhancing political accountability. Hence the argument that both increased environmental protection and reduced chances of conflict between states is the likely outcome of democratization. Here is the possibility of a virtuous circle: democracies don't usually fight each other, nor do they produce environmental destruction that might, according to the 'environmental degradation leading to conflict' argument, lead to instability and violence.

But there are a number of geopolitical arguments related to this position that need to be taken seriously in thinking about the formulation of an 'environmentally benign democratic peace'. The first apparently strengthens the case in a slightly ironic way. While democracies are, according to recent scholarship, much less likely to fight each other than authoritarian regimes, some scholars have recently suggested that the historical record of the twentieth century suggests that advanced socialist states are even less likely to fight each other (Oren and Hays, 1997). Whatever the empirical merits of this case, given the environmental record of the Soviet Union, clearly lack of bellicosity is not necessarily related to environmental performance. Industrialization is not the issue in these formulations, democracy is.

But the second geopolitical argument about the tendencies of democracies to be both peaceful and environmentally responsible, suggests that the model is missing a number of crucial dimensions. As was often noted in the period of the cold war, democracies often exported their violence, involving proxies in some nasty wars in the 'South'. Once again the argument about the democratic peace can be saved by arguing that the democracies were not supporting proxies that fought each other, but the tendency to get involved in violence in what is now termed the 'South' is important in understanding the matter of the democratic peace in the current geopolitical context. The model of a zone of peace is premised on a zone of turmoil outside.

Third, as noted above, by relying on such a formulation of the terms of understanding of the environmental security problematique, the environmentally benign democratic peace argument does not include the point that many of the resources for Northern consumption are made available at the cost of degradation in the South. This explicitly raises the questions of 'state-territorial' assumptions underlying the geopolitical units made into the objects of analysis in the discussion of the environmentally benign democratic peace.

More specifically, if the Soviet Union was, as the center of a geopolitical bloc, both an advanced industrial socialist state, and also a very environmentally destructive one, might this be because, at least in part, its resource extractions and its especially dirty industries were both within its borders? In comparison, the advanced democratic states have often acted in ways that effectively distance themselves from the worst sources of pollution and degradation. If the non-Soviet world is understood as one geopolitical bloc, the comparison might suggest that the environmentally benign peace is convincing only because the hinterlands that are despoiled are substantially removed from the democratic states by the global pattern of resource trading and 'pollution exports'.

In short, the question of the location of environmentally benign societies is in part a convenience of territorial state boundaries and the assumptions of states as the containers of both politics and environments. This example comparing blocs, rather than states, suggests that the geopolitical categories used in discussing environmental security are an important consideration that cannot be taken for granted. The exclusion of resource flows and other transboundary processes from state-territorial understandings of politics raises the issue of the history of these patterns and the longer term processes of colonization and the disruptions caused by the expansion of modernity.

Geopolitics or ecopolitics?

Much of the focus in the environmental security discussion has been on projecting current trends into the future to anticipate likely disasters and instabilities that can hopefully be avoided by adopting appropriate policies in advance. This has been usefully challenged by Thomas Homer-Dixon's (1994, 1996) insistence that environmental disruptions are *already*, in part, causing political conflict. But the questions raised by investigations of environmental history have yet to be comprehensively incorporated into the discussions of either the larger questions of environment and security, or the more focused scholarly investigation of the environmental degradation leading to political conflict hypothesis.

If questions of environmental history are addressed seriously then the geopolitical assumptions which structure the contemporary questions of the politics of global change may need to be changed. Ecological history is a relatively new scholarly enterprise, but its accounts of human induced change point to the importance of taking the long-term view and of being skeptical of contemporary claims that environmental changes offer novel political hazards. The work of historians, ecologists, geographers, paleobotanists and scholars from a variety of other disciplinary perspectives is relevant (Turner, 1990; Mannion, 1997; Smil, 1993).

More than most political scientists, Alker and Haas (1993) understood the need to think about nature in the long-term in their most useful account of the politics of the biosphere as an object of knowledge. They argue that reading Vernadsky's (1945) ideas of a single biosphere and Braudel's (1981–84) historical formulations of the macro-patterns of civilizations suggests more appropriate schemes for understanding the dimensions of global change. Alker and Haas (1993) suggest that the political framework that is invoked to think about global environmental change and the politics that are connected to this theme can be understood in terms of concerns with either geopolitics or ecopolitics.

Geopolitics is derived in part from Darwinian notions of biological competition and Malthusian-derived notions of the survival of the fittest. Applied to European rivalries and competition for territory through organicist models of states, this perspective found its apotheosis in the Nazi appropriation of Friederich Ratzel's (1897) ideas of *lebensraum* to justify political aggression. In contrast, ecopolitics is derived loosely from formulations of ecology as the study of the total relations of an organism with its surroundings, and from Verdnadsky's (1945) understanding of the global biosphere as an ecosystem at the largest scale. This recognition of the earth as a single dynamic system has subsequently been absorbed into concerns with sustainable development, articulated most obviously in the deliberations of the World Commission on Environment and Development (1987). It also appears in environmentalist thinking influenced by various schools of philosophical holism and in the recent discussions of the 'Gaia hypothesis'.[1] (Lovelock, 1988).

Alker and Haas (1993) suggest that the geopolitical vision of nature in terms of competition between states at the largest scale won out during the cold war in the formulation of the containment of the Soviet Union as the most important political task for the West. This is related to the neglect of considerations of global ecopolitics. But the spatial imaginary of cold war geopolitics, and the perceived necessity to expand control over territory in the struggle with the other bloc, is of course not only a neglect of ecopolitics. It is also crucial to the processes justifying military preparation and industrial economic growth that have been an important cause of degradation of the natural environment (Finger, 1991).

While this focus on the macro scale may not generate the immediate policy-relevant scholarship that the Marc Levys (1995) of this world desire, it does suggest that the questions posed about the politics of the global environment cannot be answered adequately by the neorealist assumptions of state-territorial actors as the key dimension of global politics and the source of the provision of adequate policy responses. The crucial point here is that the concern about the global environment requires thinking about matters that transcend both the state and the conceptual tools of contemporary neoliberal scholarship (Keohane and Levy, 1996). Notwithstanding some of its more flexible assumptions about the possibilities of social learning and regime construction, neoliberalism still operates on the state-territorial geopolitical premise that is part of the problem for those who wish to think about global problems in the environmental sphere.

Industrial imperialism

Thinking about matters in terms of changes to the biosphere, rather than as geopolitical competition, suggests that the timescale for consideration of global change should not be taken for granted any more than should geopolitical entities. In policy-oriented scholarly work, concerns about climate change are usually focused in terms no longer than a few decades. This is understandable in organizations like the Intergovernmental Panel on Climate Change (1992), which has a specific rationale in providing advice and suggestions to international organizations and governments. But such a focus may not be very useful in understanding the most important long-term trends. Ironically, this is also probably the case in circumstances where, as some recent research suggests, rapid climate change may come unexpectedly (Alley and Bender, 1998).

Global atmospheric change is driven most obviously by the widespread combustion of fossil fuels used in the processes that Lewis Mumford (1934) once so acutely called 'carboniferous capitalism'. In biospheric terms, users of fossil fuels are transferring carbon from the rocks of the earth's crust into the atmosphere, reversing a long-term natural process of sequestering carbon from the atmosphere. In the process, fossil fuel consumption is apparently disrupting the temporary stability of the global climate. That, of course, is not quite what the early pioneers of the steam engine and the factory owners of two centuries ago understood themselves to be doing. But understanding matters in terms of at least the last two centuries is necessary to deal with climate change.

Thinking back two centuries to consider the implications of environmental change takes us back to a world of Napoleon. It takes us back to before the Congress of Vienna. The United States was but a collection of former colonies that had yet to forge a unitary identity. It takes us back to a period in history where the wealth per capita around the world was loosely equitable prior to the dramatic rise in European power and affluence. The view from the end of the eighteenth century focuses attention on the dramatic changes that were to be wrought by industrial capitalism in combination with the forces of the second round of European colonial expansion. It also suggests that while European dominance was an important dimension of the geopolitical arrangements on the planet, it had not yet directly shaped large parts of the planet by its colonizing practices. India still had a textile industry, although the growing power of the coal-powered cotton industry in Britain, combined with imperial trade arrangements, was soon to put an end to it.

The point here is that the geopolitical context of the last two centuries has changed dramatically while the driving force of anthropogenic atmospheric change has emerged. The framework of contemporary politics is very different from that in the days of Napoleon but the political processes understood in the conventional neoliberal and environmental security frameworks are largely limited to those that specify politics in modern statist terms. This is not to make some suggestion of the superiority of global civil society as an overarching framework to understand global politics, nor to suggest that there is necessarily one other appropriate structure for dealing with the difficulties of climate change. It is merely to reiterate the crucial point that statist analyses and a cartographic imagination of the nation-state and policies of regime construction are unduly constrained in providing political and conceptual tools for grappling with contemporary ecopolitical difficulties (Lewis and Wigen, 1997).

Putting matters in historical context is necessary. Going back even half a century shows that the nascent political system of the United Nations of 1948, within which these matters could have been investigated then, is very different from that of 1998 when this chapter is being written. Thinking about what we need to do now to anticipate ecological conditions in 2048 or 2098, never mind 2198, makes the point about the appropriate scale for consideration. Independent states didn't exist in large parts of the world half a century ago, and so the context within which the frameworks currently being operationalized in such arrangements as the Kyoto agreement on atmospheric change didn't exist either. The problem of global change and appropriate responses is more fundamental than is often recognized.

Pre-industrial imperialism

But this question of global change, ecopolitical structures and their relevance to thinking about environmental security can perhaps be better understood if the time-scale is extended back still further. Jared Diamond's (1997) magisterial overview of human ecological history is especially instructive in understanding the macro patterns of the world that have brought us to the current concerns with global change. His suggestions lead to an understanding of the human expansion over the planet as a profoundly disruptive influence long before industrial and colonial processes accelerated changes in the last few centuries.

From the destruction of numerous large animal species by overhunting, through the subsequent emergence of crop planting and the resultant clearance of 'natural' vegetation, to the emergence of domesticated animals, specific groups of humans have long acted in ways that either deliberately or accidentally changed both their environments and their relations to other neighboring humans. The emergence of human diseases as a result of animal domestication in particular places in Eurasia had implications for the specific populations who subsequently gained partial immunity to these diseases. This gave them crucial competitive advantages over non-immune populations when they came into contact.

One of the most interesting arguments that comes from Diamond's (1997) reconstruction of ecological history is the recognition that the question of organized violence and the politics of destruction of weaker human societies is an ecological and geographical factor on the largest scale. Agricultural peoples have gradually, or sometimes not so gradually, displaced hunter-gatherer peoples. This process has obviously accelerated enormously in the last half millennium where the expansion of Europeans into the Americas literally decimated their populations, removed much of the forest cover and indigenous vegetation, and spread a new agricultural ecology across much of both continents (Crosby, 1986). Critics of the colonizing practices of Europeans and of many contemporary states have long bemoaned the destruction of native cultures, peoples and environments. The industrial imperialism of the last two centuries has merely accelerated and extended the process.

The indigenous peoples destroyed in the process have often been nearly invisible to the dominant triumphalist narratives of modernization in the 'North–West'. The expansion of European empires came at the cost of numerous indigenous societies and their destruction and obliteration, becoming, in Eric Wolf's (1982) still very apt phrase, 'the people without history'. What is important to understand is that this was often a violent matter of conquest, with the prizes for the conqueror being land and other 'resources'.

Viewed in this long-term pattern of seeking land and food in the patterns of conquest, the question of environmental security and the relations of violence to environmental change take on a very different significance. The encroachment of new ways of life into areas previously organized on different principles is usually a violent process. The relationship between violence and environmental change is thus not at all a new process. While many of the environmental security authors recognize some aspects of this lack of novelty, understanding the violence in such places as Brazil in terms of the expansion of modernity, suggests that the traditional interpretations of land wars on the agricultural frontier may explain much more about contemporary patterns of environmental change and violence than Malthusian assumptions about environmental change as a generator of conflict (Hecht and Cockburn, 1990). Seen as part of the processes of the expansion of modernity, political violence can be understood to be an intrinsic part of environmental change.

The erasure of indigenous populations is related to the amnesia in the cartographic specifications of contemporary administrative practices. As Mike Shapiro (1997) argues, the North American landscape is a conquerors' construction where the history of resistance is obliterated from maps and memories. Whether as a direct colonial administration, or through the commercialization and privatization of collective resource use, the practices of appropriating environments continue apace (Johnston, 1994; Howitt, Connell and Hirsch, 1996). If these practices are understood as the driving forces of migration, as well as environmental and political change, then one gets a view of the processes involved in the environmental security problematique that is very different from conventional assumptions that environmental change may lead to insecurity. Viewed in historical and geographical terms, the practices of violence that are of concern in the environmental security discourse are not in any way new. Rather, they are the continuation of a pattern of conquest and dispossession that has a long and bloody history.

Geopolitical imaginaries

But now these processes are also driven by the power of industrial technology which has, both directly through fossil fuel emissions, and indirectly through the dramatic expansion of rural exploitation to provide raw materials, accelerated the processes of change. Thinking of nature as separate from the administrative practices of states and the cartographies of divisions and borders misses a crucial dimension of this dynamic. As Neil Smith (1990) has shown, these are in fact better understood as two facets of the same process. A forest cleared by a new 'settler' to plant a field of crops is simultaneously a change of land use, environmental change and a spatial practice of enclosure supported by colonial land management arrangements. Space and nature are produced simultaneously. Considered in environmental terms, the division of ecosystems into separate parcels of land is a process of land use change that has dramatic environmental repercussions.

Again the historical construction of these geographical matters is crucial. The point that many 'Southern' peoples make, in arguing about the responsibility of those in the North for global warming, is that the historical record of Northern consumption is a crucial factor in the debate (Dalby, 1999). Environmental changes have already been undertaken dramatically in the temperate parts of the world that Europeans colonized. Assuming that current environmental change elsewhere is a problem, but that the historic changes wrought by colonization in North America are not to be considered, is to obscure the relevant context for discussing global environmental security. Put in this geopolitical framework, the questions of 'who is securing what, where' become especially pressing.

Diamond's (1997) analysis is also clear on the point about populations being in motion in the long-term. Assumptions of spatially stable populations make as little sense as an ontological premise for politics as they do for understanding ecological realities on a scale necessary to grapple with climate change. Over the long course of human history, populations have moved and displaced other humans and their ecological niches; there is no reason to suspect that the twenty-first century should somehow be different. Thus, a larger geopolitical vision sensitive to global change as an ongoing historical process, rather than as something new, suggests that the recent administrative boundaries of the post-colonial world are a less than useful set of categorical entities to use as the framework for an analysis of either the current problems, or the long-term directions for humanity.

Insecurity and change

But environmental discourse, as well as political discussion, is often caught in assumptions of stability or at least homeostatic equilibrium (Botkin, 1990). The important point for the contemporary literature about both development and environment is the recognition of the processes of change as the context within which these discussions must take place. The most basic concerns of the discourse on environmental change often betray their premises by posing change as a problem. This in turn assumes stability and the political *status quo* as the acceptable baseline for discussion.

The contradiction is enormous not least because development is so widely understood as providing much of the solution to current difficulties. Squaring this circle is usually a matter of talking about political stability as the crucial dimension to be assured so that development can occur. The environmental conservation assumption of stability and preservation is frequently mapped onto political discussions of order and the lack of violence to formulate environmental security. But the dominant structures of the global economy, in its (violent) processes of expansion and disruption, are premised on change. The biosphere itself is also changing and ecosystems are only at best temporarily in a homeostatic state.

At the largest scale, the assumptions about biological conservation obviously make sense in that changing the composition of the atmosphere may induce unpredictable rapid changes in the climate. Likewise radically reducing the overall biodiversity of planet earth, and in the process reversing the very long-term trends of evolution towards greater diversification, is also a dangerous gamble given the adaptable flexibility inherent in diversity. But these are not necessarily a useful set of criteria to directly apply to the smaller-scale considerations where disruptions and migrations are already in motion.

Beyond these difficulties, formulations of environmental security often fall back on the statist arguments that suggest that states are the only possible innovative mechanisms and the providers of security, so they must be the most likely method of doing something useful about limiting environmental damage. But this leads back to Matthias Finger's (1991) argument that the environmental security problem itself is driven substantially by states' attempts to render themselves secure in an industrialized world of security dilemmas. They do so by appropriating natural resources to facilitate the industrial production of military systems.

The alternative suggestions of taking global civil society seriously suggests the necessity of thinking about politics in more complex ways and understanding the potential of useful responses as originating outside the formal structures of states. (Wapner, 1996;

Lipschutz, 1997). Here at least the dynamics of change are more clearly in focus, but as a panacea for political action these proposals are also obviously limited, especially when they link to political strategies based on uncritically accepted spatial assumptions of the need to protect 'local' communities from 'external' interference. (Stewart, 1997) Nonetheless, they are at least closer to a framework for understanding the ecopolitical processes of change than more conventional geopolitical modes. Not least, this is because these frameworks are more explicitly skeptical of both the claims of states and the dynamics of development.

Ecopolitical imaginaries

Environmentalists have frequently argued that there is a fundamental ontological mismatch between states, their practices of sovereignty, and the ecological crisis (Conca, 1994). That much is not a new argument. But the crucial points in this chapter about the historical dimensions of environmental change as a practice of violence, and the geographical understanding of the historical mobility of people as anathema to the assumed political containers of the contemporary international system, suggests that the current debate over environment and conflict is caught in a series of disciplinary and institutional limitations that fail to grapple with either the scale of the problem or its historical trajectory.

The difficulties of understanding both politics and ecology as processes in motion, rather than as stable entities in need of securing in the face of change, are considerable. But any substantial engagement with the discussions of the environmental security problematique has to take these matters seriously. (Dalby, 1998b) Specifically, the dangers of environmental determinism, and neo-Malthusian assumptions about production limits in particular environments, need to be countered by a more sophisticated political ecology that understands environmental change as a series of complex social processes in specific geographical contexts. In the process, simplistic assumptions about the efficacy of states, markets, civil societies or projects of ecological modernization to provide solutions to environmental difficulties are put in doubt. (Muldavin, 1997)

Environmental change simply may not be the most important variable with which to start an analysis of environmental security. If environment is understood as a cultural process in the long-term, security becomes a much more difficult term to invoke as an overarching policy objective. Focusing on migration and ecological disruption as the human condition, and on the dispossession of less powerful peoples as a long-term process accelerated and extended in the history of European expansion, carboniferous industrialization, and more recently, contemporary globalization, offers a very different history and causal sequence for understanding environmental insecurity.

As a result, in addition to reaffirming the obvious importance of thinking critically about the spatial limitations of conventional international relations writing (Walker, 1993), the analysis in this chapter suggests four specific responses to contemporary formulations of environmental security. First, the questions of environmental security are new only in the sense that the post-cold war political situation allows for them to be discussed in a manner that the geopolitical rivalry precluded. Ecopolitical matters are on the agenda, but conflict and environmental change have been intimately connected in the expansion of modernity for centuries.

Second, the environmentally benign democratic peace, however desirable it may be considered to be as a policy goal, is premised on a (geo)political ecology that needs to be explicitly worked into the model. Questions of resource production, use and trade, and the specific geographies of environmental degradation, complicate the state territorial framework of such theorizing. Incorporating these matters into the analysis also suggests, third, that the links between inter- and intra-state conflict in the analysis of environmental security need further elaboration in ways that incorporate environmental change directly into such models.

Fourth, the whole question of the geopolitical premises in such analyses cannot be ignored if the theoretical analysis is to be robust enough to incorporate both matters of environmental change, and the politics of security in ways that grapple with the sources of environmental change at appropriate scales and in adequate complexity. Ecopolitical considerations require that ecology and environmental history be taken seriously (Dalby, 1998b). While decisions about humanity's future are clearly political questions, the contextualizations in which they are thought about, debated and decided need much more careful attention than has so far been the case in most discussions of environmental security.

Notes

1 A much earlier version of this chapter was presented as a paper titled 'Environmental Geopolitics' to a panel on 'Facing Change: Environment, Sovereignty, Security' at the annual meeting of the International Studies Association in Minneapolis in March 1998. My thanks to the participants in Minneapolis, to Miriam Lowi for editorial suggestions and to Cara Stewart for discussion of environment, space and politics. The original paper was prepared under the auspices of a Carleton University Research Award for research on 'A Critical Geopolitics of Environmental Security'.
2 The Gaia hypothesis suggests that the earth's biosphere is a self-regulating system that has maintained the key conditions for the preservation of life on the planet despite long-term changes in solar radiation and major geological events.

Part II
Practices

7
Integration of Non-Traditional Security Issues: a Preliminary Application to South Korea

Robert E. Bedeski

Introduction: the notion of Non-Traditional Security

Values

In the twentieth century, the state itself has become a major source of social and political values – especially for developing countries. Democracy, justice, equality and other values of the modern industrial state are part of the vision that inspires Third World leaders, but most developing countries see sovereignty and independence as nation-states as either recent acquisitions or yet incomplete goals. The project of state-building thus is particularly sensitive when the topic is cooperation or modification of autonomy. The US and other industrial states often demand concessions and compromise from developing countries, which are still at a nineteenth century level of state-building. Part of this conflict of vision – complete versus incomplete states – comes from the relationship between security and freedom, in that developing countries see liberal institutions of democracy more as post-industrial luxuries than as prerequisites to full industrialization. How can we build the mansion of individual liberties demanded by the US, the countries of the Third World ask, when we live in a hut of poverty and illiteracy? The Western reply is that prosperity and democracy are mutually reinforcing, but historical proof is difficult to find. In the case study of South Korea, development preceded democracy for more than a generation, and a longer period of gestation was needed in pre-war Japan, and had to be re-established after World War II.[1]

Now that the cold war is over, the world can turn to the accumulated problems generated by industrialization, demographic explosion, and military conflict. The global community can use technology, international cooperation, state institutions and even market solutions to ameliorate and head off a range of Non-Traditional Security problems.

Neither the industrialized nor the developing countries have the luxury of inventing new technology and institutions to deal with these issues, and so the challenge is to adapt existing science and structures to confront these threats. The modern state has evolved specific institutions which can be utilized for certain aspects of security. If existing institutions are sufficiently flexible, they can be adapted to the new challenges. The use of military forces in the drug war and interdiction of illegal immigration are examples of this kind of adaptation. Likewise, the use of space satellite imaging in monitoring environmental problems is another area of what was previously national defense technology becoming a key tool in Non-Traditional Security.

Non-Traditional Security addresses the well-being of existing states and their citizens and the growing responsibility of states to their inhabitants and to a larger international community. There is growing awareness that modern states can preserve themselves only by cooperating with other nations and by taking actions only in concert with the international community. Unlike *Traditional Security*, which narrowly focuses on military protection of the state and its territory, *Non-Traditional Security* proceeds from the observation that states are threatened by a broad range of challenges – some of which are generated within state and society. There is also implicit recognition that the nation-state has been the major unit of social and political development during the modern period, as well as the basic unit of international relations. It is not likely to be replaced by any global organization, or regional groupings. The modern nation-state has serious flaws but is the universally accepted form of political organization in the world today.

'Ongoing systematic changes in the world today have given rise to new, non-military sources of instability... for example, extreme poverty, political upheaval, refugees and the systematic suppression of human rights.' (Japan Times Weekly (International Edition), 1993 April 26–May 2, 3). Stoyan Ganev, speaking at the first plenary session of the Third United Nations Conference on Disarmament Issues in Kyoto, called for a new comprehensive approach to security, and said, 'We must address the security concerns of nations within the framework of economic and social development... and protection of human rights.'

Non-Traditional Security seeks to identify the non-military challenges to the proper functioning of modern nation-states, and elicit responses appropriate to maintaining gains in well-being without creating new threats to external citizens and states, or future generations who will have to bear the costs of current prescriptions and cures. Another element of Non-Traditional Security is that resolution of non-military threats must rely on more international cooperation rather than confrontation. Dialogue and the search for cooperative solutions is increasingly warranted by the transnational character of Non-Traditional Security challenges. Although this manner of bilateral or multilateral approach appears to infringe on sovereignty, it proceeds from a calculus of self-interest and is no more a threat to sovereignty than are treaties.

Non-Traditional Security was introduced into the North Pacific Cooperative Security Dialogue (NPCSD), initiated by Canada in 1990, in order to broaden the traditional concept of security. Traditional Security generally refers to the project of protecting states, groups and individuals from military threats. The threshold of force and violence is usually clear, while national and international institutions for dealing with these threats have been articulated over centuries, with frequent updating. Non-Traditional Security, on the other hand, has no similar clarity of definition, nor specific institutions to maintain it. Yet we are increasingly aware of a growing gray area of threats to national well-being, which has no direct military component, no adequate military response, and few functional organizations designed to deal with these threats.

We can suggest a working definition of Non-Traditional Security: those non-military issues that disturb national well-being, regional tranquility and international peace. As we subdue the military threats through arms control, CBMs, and multilateral diplomacy, we can expect non-military threats to play a wider role as sources of international friction. An important task is to identify these potential problems and suggest directions for future policies that governments can pursue. International euphoria over the passing of the cold war was short-lived, as it was replaced by the cold reality of ethnic nationalism. The era of mutually assured destruction (MAD) has been replaced by new insecurities, including the breakup of the Soviet Union, economic decline in some regions, new waves of population migrations, and growing environmental crises, to name a few. The Soviet–American standoff had postponed attention to a wide range of non-military problems.

The idea of Non-Traditional Security can provide insights which have not been addressed adequately. The concept proceeds from the observation that threats to national security can originate from sources other than military. Traditional Security threats are characterized by several criteria:

- They usually involve the use of military forces with the use or threat of coercion and weapons.
- They are usually explicit – consisting of belligerent acts, or activities, which increase the likelihood of belligerent acts, or succeeding in these acts. They include activities such as military attacks, military buildup or movements, or alliances.
- They are usually intentional, and governments commit resources to acts that have the force of the state behind them.
- They generally have recognizable goals, which when achieved, will result in some reorganization of the threat mechanism. After successful invasion, for example, an army of occupation will replace the army of attack.
- The resolution of the threat will usually involve a redefinition of relations between two or more states.

In contrast, Non-Traditional Security threats,

- arise out of the normal, non-military activities of individuals, groups and states;
- are cumulative, and usually not perceived as threats, at least in the short run;
- affect the ability of governments to maintain stability and enhance living conditions within state borders;
- and are neither explicit nor purposeful threats of governments.

It is this last characteristic that makes Traditional threats difficult to identify and resolve. There is the further obstacle to resolution: At what point does a Non-Traditional Security problem become a threat requiring significant expenditure of national wealth and resources? Furthermore, when either, occur, how will responsibility be assessed?

The major difference between Traditional and Non-Traditional Security threats is that the latter are generated more by the daily mundane activities of living, working, travelling, interacting and procreating. These activities are increasingly observed and measured by governments, but generally not controlled by the state. The ordinary quality of these human activities makes perception and definition difficult. Human activity has been dangerously successful in the conquest of nature, and now industrial and electronic technology amplifies the consequences of these actions. Technology allows unparalleled efficiency, acceleration of human interaction and the subduing of the natural environment. In some sectors of activity, excess is leading to a threshold where the well-being of individuals, societies and states is unraveling. Thus far, governments have been able to manage some of these changes, but we are only seeing the beginning of a new range of problems – from over-population to global warming. More international cooperation is needed to manage these Non-Traditional Security challenges.

In the following section, I examine the application of Non-Traditional Security concepts to the case of South Korea. While traditional military threats are still a grave concern to South Koreans, rapid industrialization has created a wide range of growing problems which can no longer be postponed.

South Korea: viability and sovereignty

Geography

While the Korean Peninsula has preoccupied military and defense planners and strategists for nearly a half-century, developments there have created new sets of Non-Traditional Security concerns – especially in the realm of environment. Rapid industrialization in both North and South Korea has brought pollution of air, land and water, and rising levels of production and consumption – particularly in South Korea – have accelerated pressures on food and raw materials. Industrial development has rapidly expanded South Korea's 'ecological footprint', adding a new dimension of human security concerns.

The new dimensions of environmental damage, illicit activities, demographic growth, security of food and resources, and human rights are fairly predictable for South Korea from the experiences of other developing countries, but specific geographical and historical realities make for a particular mix of problems which must be addressed. Because the Korean War (1950–53) was never ended, South Korea faced little relief (except for normalization of diplomatic and economic relations with former opponents of the People's Republic of China, the Soviet Union, and the other allies of North Korea) from the ending of the Cold War, and could not transfer attention and resources from military to non-military application.

Seoul must deal with immediate threats to survival from its only land neighbor – the Democratic People's Republic of Korea – and with Non-Traditional Security challenges arising from successful industrialization simultaneously. In addition, South Korea is located at the vortex of Chinese, Russian, and Japanese interests in the region.

This geopolitical burden has resulted in repeated wars and occupation by foreign powers in the past century. South Korean economic growth has been linked to that of Japan, and to trade with the US and its allies, and now is increasingly tied to the rising growth of China and the openness of the Russian Far East. Until 1997, the economic miracle appeared to be self-sustaining, but late in the year, the so-called 'Asian financial flu' attacked and injured South Korea, causing bankruptcies and unemployment unprecedented in decades. At the same time, the North Korean regime has suffered years of negative growth and recent famines, generating amazement that the system has not yet collapsed. South Korea will bear the major brunt of any North Korean collapse, and must prepare for what seems to be inevitable.

For South Korea, the challenges to security are hinged to its very survival. This raises the issue of priorities for the state – economic growth through industrialization and exports increased the complexity of security planning, which had been largely military through the 1960s, when President Park Chung Hee embarked on his developmental strategy. Now South Korea is part of the global economy, and its political, social, economic, military, diplomatic and environmental policies must be coordinated accordingly. Preserving survival remains the state priority, with reliance on military force at the core – as it is in North Korea – while enhancing wealth of society has become the second priority. The third priority is Human Security, or Non-Traditional Security. The problem for South Korea is that its historical/geographical vulnerability sharpens practically every military or economic crisis to an issue of survival, with the result of pushing aside the issues of Non-Traditional Security. During the administration of President Kim Young Sam (February 25, 1993– February 24, 1998) Non-Traditional Security issues were given more attention than ever, but the new President Kim Dae Jung faces severe economic crisis as well as the smoldering obstreperousness and potential collapse and/or violence from Pyongyang. While the two Koreas do not face internal ethnic strife due to their high degree of homogeneity, their mutual distrust and anxiety is as high as that between Israelis and Palestinians.

South Korea is a paradigm of developing countries in at least one way: late development is a symptom of weak state institutions, or delayed achievement of national sovereignty. Achievement of high economic performance is tentative and vulnerable – as even the Japanese have discovered. This means that state survival will take precedence over all other considerations of security; it is for this reason that the modern sovereign nation-state – not multinational corporations, NGOs, international organizations, or technological breakthroughs remains at the center of any analysis of security; military or non-military.

Today, South Korea stands at a historical crossroads of its development as a nation-state. Its people have survived and overcome the numerous challenges of war, reconstruction, economic development, and democratic transformation. Some of this success can be attributed to fortune, but several important other factors have also intervened; in particular, there has been an open society, a market economy, pluralism, and a core of strong institutions – especially the presidency – which have allowed the country to move forward. Some of the success was due to a successful gamble – and here I would include the Nordpolitik, which took advantage of Soviet collapse. The South Koreans have engaged in rapid modernization without abandoning elements of traditional Confucianism, which have reinforced transition. Continued Korean survival and growth depends on mastering the skills of modernity on the one hand, and preserving the attributes that have served the country well in the past. Mastering modernity, however, does not mean wholesale imitation of the advanced industrial nations, although it does require increased integration into their patterns and economies.

Military, or Traditional, security is relatively easy to identify and to assign to a sector of state activity. The Non-Traditional Security challenges, however, are far less institution-specific. Sustaining of economic growth involves a mix of private and public sectors. Banks, corporations, labor unions and government ministries all play their roles. Moreover, this area of security does not have clear-cut definitions of allies and friends, and ambiguities abound. Labor reform in South Korea, for example, has provided better conditions for workers, but has also slowed economic growth as companies seek cheaper labor inputs abroad. Lower growth, but a more equitable distribution of wealth, may be the result in the short run, but political stability can be a significant reward over the longer run.

Economic security

A persistent belief has been that poverty and deprivation are root causes of social and international conflict. Numerous cures and theories have been proposed, with two major lines of argument underlying them. The first is that wealth is relatively fixed, and the antidote to poverty must be redistribution to the deprived sectors of society and the world. Theories of dependence proceed from this assumption. International aid has often been framed in these terms. A more radical approach borrows from anarchist thought – that all property is theft, and redistribution (including ODA, or Overseas Development Assistance) is merely restitution. Foreign aid is not altruism, but self-interest. A variation of the argument is that misallocation of wealth to arms has also been a source of deprivation. This includes not only military hardware, but also, that the allocation of human talent to non-productive military procurement programs has slowed economic development. Low defense spending in Japan has not hurt its economic expansion since World War II. In this line, reallocation of the 'peace dividend' should provide for an increase in usable wealth in the US and the former Soviet Union.

The second line of argument is that wealth is created through a combination of labor, capital and technology. It is presumed that these are available in all societies in different mixes. Supply-siders look at the record of the past century, with the phenomenal growth of gross wealth, and see evidence supporting this hypothesis.

In the final decade of the twentieth century, we have considerable evidence of what works and what doesn't work in the expansion of wealth. To oversimplify, central planning, economic autarky and suffocating state intervention interfere with growing prosperity. What have worked are export-oriented industrialization, market economies, flexible adaptation, and some degree of intelligent government coordination. There has also been a high degree of correlation between respect for private property and freedom, and successful development (De Soto, 1989). In the twilight of twentieth-century socialism, economies with large public sectors are increasingly uncompetitive.

The techniques of economic development are not a mystery. But we may only be dimly aware of the threatening consequences of its widening and uneven success. In the North Pacific Region, we find some of the most dynamic economies of the world. Japan and the US lead the world in economic production, while Taiwan, South

Korea, Singapore, and Hong Kong have grown rapidly in the past several decades, while socialist economies have grown slowly or sporadically at best, and face significant challenges in revitalizing themselves. Should these trends continue, the region will be divided into 'have' and 'have-less' countries. The 'advanced nations' have some responsibility to aid the developing societies to escape the poverty which often creates hostility towards the more well-off. China opened its doors in the past decade to allow significant foreign investment, while the Russian Republic is undergoing traumatic political reconstituting in the wake of the demise of the USSR; Canada has played a role in development assistance to China in the past, but reduced this aid in the wake of the Tiananmen Square incident.

Through practice, experiment, analysis and failure, we are becoming adept at expanding wealth without relying on radical redistribution. In the Pacific Rim, we see some of the most successful economic miracles. The market-driven, export-oriented economies require an international regime of free trade and political liberalization to maintain their economic growth.

A question we will have to address is how uneven development in the North Pacific region may lead to instability and conflict, and what steps should be taken to avoid the negative scenarios. With the end of the cold war, we also have a historical opportunity to shift arms spending to peaceful development.

Speedy economic growth occurred during the Third Republic and Fourth Republics of South Korea, with interdependence between the military and business. Big business relied upon political favors for expansion and contributed to election campaigns. In return they sought rewards in low interest loans, tax breaks, liberal export quotas and protective measures (Kang, 1988). The effectiveness of this arrangement was demonstrated in high growth rates although accompanied by several scandals. President Park Chung Hee pursued economic development as a factor to strengthen national security. Poverty and dependency threatened national sovereignty as much as the military menace from the north. Subsequent administrations have pursued economic security with the same vigor as Park (Healey and Lutkenhorst, 1989).

However, the generation of domestic wealth in an interconnected world increasingly carries with it a responsibility to assist less-developed countries – if for no other reason than self-interest. Poverty creates instability and resentment, which erodes international peace.

Trading partners will buy more products and services if they are well-off rather than desperately poor. The Japanese experience in foreign aid is an example of combining this self-interest with altruism, especially in Southeast Asia (Orr, 1987). Another characteristic of South Korean economic development has been the control that government has exerted over foreign economic inflows, minimizing the level of foreign ownership and market control (Mardon, 1990). This has helped to maintain economic sovereignty and avoid the dependency relationships experienced by other developing countries. At the same time, the government provided incentives for foreign investment, which played a significant role in economic expansion (Stoever, 1986). Recently, South Korea has reversed its role from recipient to donor of aid; with particular emphasis on Russia.

Environmental problems

The combination of economic growth, population growth, and technology is creating environmental degradation locally and on a global scale. Ozone depletion and global warming, if dire predictions materialize, will affect everyone everywhere, and will change weather patterns in unknown ways. Oceanic pollution has increased rapidly with industrialization, urbanization and rising use of chemical fertilizers. This will endanger the continental shelf waters that are often the richest in sea-life.

Technology has enabled us to use the natural environment as a mine and a kitchen sink. It is a mine in the sense that we draw our valuable resources for manufacturing, energy and food from it, while putting little back in return. These rich resources are the accumulation of millions of years of evolution, but are not inexhaustible. The environment is also used as a kitchen sink when we use it to discard and wash away the byproducts of our agricultural and industrial activities. Burned hydrocarbons, acidic effluent, urban sewage and chemical fertilizers are a few of the products which end up on land, and in the air and water. With increasing industrialization and burgeoning populations, environmental degradation and destruction will accelerate.

Atmospheric pollution is a serious problem in numerous cities on the Pacific Rim. Ocean currents and winds carry these pollutants far afield. Chernobyl's radioactivity in the Ukraine went not only to Eastern Europe, but was detected in Western Europe as well. These kinds of problems will only grow larger, and will require transnational efforts. We increasingly recognize the environment as an indirect instrument of war – from the US use of Agent Orange in Vietnam to Iraq's torching of the oil fields in Kuwait. These are Non-Traditional Security threats originating in traditional acts of war.

Korea's environmental security

A second area of security for South Korea involves environmental protection. The major dilemma for the environment is the conflict between short-term and long-term rationality. In the short run, the natural environment is treated as a passive and silent partner in development. Land, air and water are treated as if they are free goods. Pollution and other forms of environmental degradation are products of industrialization, and South Korea has certainly witnessed these phenomena. The costs of unrestrained industrialization are already adding up, and future generations will have to pay an increasing price for the industrial accomplishments of present society.

Each nation's geography and level of economic development profoundly affect its environmental problems and opportunities. Korea is characterized by poverty in natural resources and high population densities (427 per sq km). Only 23 per cent of land area consists of lowlands and plains – the type needed for urban, industrial, and agricultural development – and most of these lie in coastal and river areas. But as rapid industrial growth occurred, it had a deleterious effect on coastal zones. Tanker traffic has brought oil spills (858 between 1983 and 1987) (Hong, 1991). Coastal waters are also damaged by urban and industrial waste dumped into rivers – estimated at 1 610 000 tons of waste water daily into the four major rivers: the Han, Kum, Yongsan, and Naktong.

In 1990, South Korea saw its first Green Party, the Korea Green Party, which began to pressure government to tighten up environmental law. Seoul set up the world's first environmental police force in 1991, empowering officials in charge of environmental protection to exercise judicial powers over persons and companies discharging pollutants illegally (*South* 1991). Nevertheless, pollution control has a poor record. In 1992, The *Korea Times* reported an increase of 83 per cent in environmental crimes, which it attributed to industrial companies taking advantage of possible government laxity in an election year (*Daily Report: East* Asia [here after DR:EA], 1992 May 18, 28).

Japan's concern with airborne pollution from China has prompted

Tokyo to propose creation of 'model factories' there to reduce gas emissions. Under the MITI plan, Japan will provide factories with desulfurization equipment and fluidized-bed boilers to reduce nitrogen oxide, and will also help to train operators in the anti-pollution machinery (*Japan Times Weekly* 1993 June 14–20, 2). Japan and South Korea have agreed on a final draft of an ecological agreement to exchange information on policies and technology, and to promote joint research on bilateral concerns related to water, air, marine and soil pollution. The two will also cooperate on issues of biodiversity and global warming. With normalization of relations between China and South Korea, and increasing investment, it will be in Seoul's best interest to cooperate with Beijing on similar projects. In October 1992, Korea had the highest reported methane concentration in the world (1823.3 parts per billion over South Korea versus a global average of 1660). Its source was believed to be China, and is related to increasing population. Methane is one of the major causes of global warming, along with carbon dioxide, which has also been drifting from China. On a global level, annual production of methane is estimated at 500 million tons, and is growing at 10 per cent per year (Seoul *The Korea Times*, 1992 October 8, 3, in *DR:EA*, 1992 October 8, 23).

An even more direct concern is pollution in North Korea, which is one of the few countries in the world where objective data on the environment is unavailable. Even basic demographic data remain inconsistent and often unreliable. Serious pollution in the lower reaches of the Tumen River and acidic rain is reported by foreigners who visit North Korea. Mercury contamination, red tides, and other indicators of marine and atmospheric pollution have been reported (Seoul *Maeil Kyongje Sinmun*, 1992 June 13, 9, in DR:EA, 1992 August 4, 34). North-South cooperation on environmental issues must be a major priority in any economic aid or investment program.

Resource security

Even in our much-touted 'Information Age', material resources – especially fossil fuels, food supplies and manufacturing substances – cannot be foregone. Populations need food, clothes, fuel and the multitude of items required to sustain survival and growth. Simple possession of raw materials does not guarantee economic development. Industrialization requires access to resources. International trade and the modern revolution in transportation including containerization – have facilitated development where few natural resources exist. Transnational usage of resources has been expanding for years, but resistance to overseas appetites will emerge as societies claim their dwindling resources for their own industrial uses. Technology continues to discover new uses for old and new resources. Previous waste products find a place in the new industrial economy as substitutes for traditional resources. One example is the western yew tree, which grows in the western US and Canada and was previously burned away in logging operations. Its bark contains a vital ingredient for manufacturing taxol, used in a newly discovered cancer drug. The tree has now become a valuable resource, and is now threatened by bark-poaching, rather than by logging operations.

The working assumption of resource acquisition in the post-World War II period has been that market forces will assure steady supplies at reasonable prices. However, the vulnerability of petroleum supplies to the OPEC cartel was demonstrated in the oil crisis of the early 1970s. Oil is one of many resources underlying industrial economies. Until other renewable sources of energy are fully developed, resource dependency will remain a fact of life. For the industrial nations, strategic interference in the flow of oil or other vital materials may constitute an act of belligerence. International dependence is a fact of life, and the depletion and exhaustion of resources will create threats to the well-being and existence of modern societies. Manufacturing, technology and trade are the media that transform natural resources into usable goods. Education is vital to allow a society to adapt, implement and continuously upgrade its technology. As population increases and living standards improve, consumption of all resources will go up. At some point, primary and secondary resources will become scarce and even depleted. The market mechanism will allow adjustments, and technology may produce substitutes, but disputes will be inevitable and their consequences, significant.

Food is a special category of resource, and each country demands food security – even at the cost of subsidizing inefficient agriculture in some cases. For the nations of the North Pacific Rim, fisheries are vital in providing a part of the daily diet of their populations. The maritime environment provides a commons where adjoining countries exploit the rich marine life. An international regime has regulated much of the fishing activity, but as in any area designated as 'commons', there is a constant temptation for users to, officially or unofficially, extend their yield. In the Pacific fisheries, this has been done by increasing the number of ships, their sizes and their efficiency, the season, and by using drift nets. Thus far, international pressure has been fairly effective in restraining overfishing, but population pressures are certain to increase the maritime burden.

Korea's food security

Food self-sufficiency has been a goal of Korean governments, with around 18 per cent of investment funds devoted to agriculture since the early 1970s. Large-scale reclamation of tidal zones has been undertaken on the West Coast, and irrigation systems have been improved in order to expand production.

According to Young Whan Kihl and Dong Suh Bark,

> In retrospect, it may be seen that the South Korean strategy of 'industrialization first and agricultural development later' has achieved its intended effect. Although agriculture also scored some gains, with an average annual growth rate of 4.8% during the period of 1963-75, the relative neglect of the agricultural sector of the economy naturally led to several major economic and political problems, which the Korean regime had to tackle. Among the major problems plaguing agricultural development in the rapidly growing economy of South Korea were (1) a widening food gap, (2) an increasing foreign exchange expenditure on food imports, and (3) a growing income disparity between urban and rural households. (Kihl and Bark, 1989, 51)

Since the early 1960s, South Korea has been a food deficit country. To reduce this deficit, the nation has relied on maritime resources. Fishing contributes less than 2 per cent of South Korean GNP, but provides 60 per cent of protein in the national diet.

Energy productivity

Korea and the other developing countries of Asia maintained high rates of growth despite oil price shocks in 1973 and 1979. Energy costs are a drain on scarce foreign currency, and an incentive to conserve and improve efficiency. The sharp rise in external debt in South Korea and other Asian states in the 1970s and early 1980s was partially attributable to the 'impact of higher oil prices coupled with high rates of income growth and derived demand for energy.' (Chern and James, 1988). While Korea has been pursuing alternative sources of energy – especially nuclear power – the economy remains increasingly vulnerable to interruptions of its oil supplies. Access to Chinese, Russian, and Indonesian sources helps to reduce dependence on the Middle East. With its long coastline, the possibility of tidal power is great, but relatively unexplored (Hong, 1991, 401).

Human rights

Why include human rights in considerations of Non-Traditional Security? Individuals are the foundation of human society; to omit them is not only incomplete but dooms discourse to failure. The study of international relations stresses national and collective entities, often at the expense of individuals who comprise the state. It is routine to link 'national' and 'security' – as if the well-being of the sovereign state addresses the major question underlying international peace. But if states have rights to non-interference from other states, shouldn't individuals have rights against undue state interference, as well? This has been the basic proposition of democratic thought since the European Enlightenment; it was reinforced in modern refinement of the liberal state, and expanded with the industrial revolution. Moreover, individual rights vis-à-vis the state are the underlying premise of international agreements on human rights, which most countries have signed.

Indeed, a nation's security can have a firm foundation only when the security of individuals is recognized and protected. This has both economic and legal–political dimensions, and has pragmatic consequences as well. When citizens perceive that their rights are protected, they will not seek emigration as the only means of achieving dignity and prosperity; instead, they will remain at home to contribute to the wealth and welfare of their own country.

Human rights include recognition of individual and minority rights, as well. In Eastern Europe and the former Soviet Union, past repression of dissent and ethnic identity has contributed to explosive violence when the old order dissolves. The era of the sovereign state legally immune to external influence and pressures – has passed, if it ever existed. States, through treaties and conventions, have voluntarily accepted restrictions on their domestic activities. It is in the interest of international community to advance standards of citizenship; in the reciprocal rights and duties of governments and individuals.

Despite advances in broader democratic institutions in the Sixth

Republic, a pattern of violations continued. According to a report of the Korean Bar Association in 1992, the 'average daily number of persons arrested in relation to political offenses was 5.96 between January and November 1990, which is far more than the daily average of 1.61 during the Fifth Republic. More than 30 per cent of those who were arrested allegedly violated the national security law, which is also far more than the 13 and 14 per cents occurring during 1988 and 1989, respectively.' ('Bar Association Issues Human Rights Report' *The Korea Times*, 1990 February 24, 2, in DR:EA, 1990 March 9, 29). Use of police torture has also aroused international outcry, while other forms of abuse of state power point to a context of violence and ideological rectitude, fueled by the tragic division of the country (*Hanguk Ilbo* 1989 October 18, 2, in *DR*:19, *EA* October 1989, 38–9) The UN's Human Rights Committee, Korean dissident groups, and others have focused on the National Security Law as a source of abuses, and urged its modification. On the positive side, however, Korea has joined three UN pacts on human rights, under which the government is obligated to submit reports to the UN secretary-general (*Yonhap* 1990 July 10, in *DR:EA*, 1990 July 10, 35).

As Western democracies insist on universal standards of human rights, South Korea can point to one mitigating circumstance: the dismal human rights record of North Korea and its continued military and subversive threat to South Korea. This does not excuse abuses in the south, but certain legal luxuries of Western democracies may have to be postponed. A report on North Korea's 12 concentration camps estimates an inmate population of 200 000 political prisoners, or close to an astounding 1 per cent of the total population.[2] (Yonhap, 1992 October 14, in *DR:EA* 1992 October 14, 27).

At the World Conference on Human Rights held in Vienna (14–25 June 1993), an alternative view was expressed, raising tensions between East and West. Some Asian governments accused the US and the West of imposing alien values derived from 'post-Renaissance liberal Western traditions'. (*Far Eastern Economic Review*, 1993 June 17, 16). At a meeting in Bangkok, a number of Asian governments insisted on non-interference, and respect for cultural particularities and historical backgrounds. China and other developing countries want human rights to include economic and welfare rights, while the poorest nations demand rights (or entitlements) to foreign aid.[3]

In this debate, South Korea is best served by adherence to the Western position because its progress in broadening human rights gives it the moral high ground – especially in relation to North Korea. Regarding economic entitlements as a component of the human rights package, it will place new burdens on South Korea's economic growth, as the poorest nations would claim assistance as a matter of right – without internal reforms or rational economic policy. Citizens will have a greater stake in the state if there is the perception that their economic conditions are improved through participation.

Institutional viability

By the end of the twentieth century, international consensus has emerged in favor of democracy as most compatible, among competing ideologies, with economic growth. With the collapse of the USSR, communism and socialism have lost much of their appeal and credibility. The major question for developing countries is no longer 'whether to move toward democracy', but 'how?' Broad institutions of representation and participation, combined with market economy, appear to provide conditions of growth and equity.

South Korea has experienced a slow and turbulent evolution of democracy since the First Republic (1948–60). Under authoritarian presidents allied with the US and challenged by a totalitarian North Korea, the Republic of Korea had sufficient stability to develop a flourishing market economy and viable national sovereignty. Following Roh Tae Woo's proclamation of June 29, 1987, the country has moved steadily to civilian democracy, reinforced by the election of an opposition party President, Kim Dae Jung, in 1997.

Illicit activities

The notion of Non-Traditional Security also includes illicit activities – narcotics trade, smuggling, piracy and terrorism. These ventures flourish where normal legal and police powers are weakest, although they may also be tolerated by some states as long as they operate underground. More the concern of police than of military forces, these activities are often transnational, and challenge the authority and ability of governments to maintain order. In some societies, these activities flourish to the extent that their practitioners constitute a state within a state. The drug cartels in Colombia, the 'Shining Path' guerrillas in Peru, the yakuza in Japan, gangs in Canada and the US ignore or challenge state authority, and the Mafia in

Italy are examples of powerful organizations engaging in illicit activities that are debilitating to millions of individuals, and drain the wealth of societies.

Such organizations have the ability to generate wealth and attract support while preying on human weakness and poverty. The flourishing of illicit activities is often a symptom of the inability or unwillingness of states to challenge these organizations. State-sponsored terrorism exists as a special case of these activities, and an indirect instrument of foreign policy. Kidnapping of hostages, aircraft bombing and hijackings have been the more dramatic enterprises, sometimes intended to attract media attention as much as to promote particular political goals. Piracy and smuggling, although perhaps less spectacular, are equally costly to individual rights, state power and national productivity.

Among Non-Traditional Security threats, these illicit activities *appear* to be the easiest to resolve, because more policing and low-level use of military forces can make a difference. But the sophisticated structure of some organizations, penetrating and influencing government up to the highest levels, and urban based, or enjoying widespread rural support, will increase the difficulty of removing these threats.

Corruption

In South Korea, there has been relatively little organized crime, piracy, and drug cartel organization. Rather, there has been a recurring pattern of official and private corruption, met by occasional government drives to clean up.[4] Moreover, the extensive security apparatus, procedures, and regulations, in South Korea obviates many potential terrorist threats – notably those of radical students or North Korean agents. Precautions often impinge on Western standards of civil liberties, but periodic incidents – such as airliner bombs or assassinations demonstrate the real and continued threat. US military or information facilities are often favored targets of attack ('Increased Security Ordered for U.S. Facilities' Yonhap, 1991 January 17, in *DR:EA*, 1991 January 17, 30). The US alliance also makes South Korea a target of anti-American states.

Nonetheless, North Korea still remains the main source of terrorist attacks.[5] Counterterrorism during the 1988 Olympics prevented any outbreak designed to interfere with the activities (see 'Olympic Security Chief on Preparedness', The *Korea Herald*, 1988 May 5, 11, in *DR:EA* 1988 May 5, pp. 21–22).

Conclusion

The Republic of Korea stands at a historic turning point in the last decade of the twentieth century. A civilian democracy is firmly established, a consistent pattern of economic growth is evident, a pluralistic society is emerging, and the main antagonist is on the brink of bankruptcy and possibly a crisis of existence. However, new and much more complex challenges replace old ones. The notion of Non-Traditional Security attempts to identify some of these as a starting point to consider how a successful modern society may see its well being and security erode.

The common feature in these threats is that they are linked to one another in different ways. In addition, each threat, or bundle of threats, has both domestic and transnational dimensions. If the North Korean military threat recedes, or becomes a smaller part of South Korea's external menace, these Non-Traditional Security considerations must be understood, analyzed and provided for contingencies.

The immediate military threats from North Korea require that South Korea focus attention on traditional defense calculations, but industrialization has added a new complex of concerns which can no longer be ignored. To their credit, the governments of the Sixth Republic (1987–) have moved on these, and can do much more in the future. If Seoul can convince Pyongyang to agree to Arms Control and Verification measures, both sides could reduce the resources expended on weapons. In a situation of mutual accommodation, North and South Korea would cooperate. In early 1998, however, with North Korea suffering economic decline and massive famine, and South Korea experiencing major financial crisis, continued confrontation appears mutually ruinous.

Other forces work for resolution of the conflict, including the Korean Peninsula Development Organization (KEDO), the four-power talks, and even the IMF. Initiating a cooperative relationship will inevitably require an Arms Control and Verification regime – unless North Korea collapses and vanishes from the map. Demilitarization of the peninsular hostilities could then move to North and South collaboration on the Non-Traditional Security challenges of the Korean Peninsula, which will be even more critical in the North in economic, environmental and food security terms.

Broadening the case of South Korea's Non-Traditional Security challenges, we see that in general terms, the challenge of Non-

Traditional Security issues is to identify them, locate their sources, and then design and implement solutions. Unlike military security challenges, Non-Traditional Security problems require cooperative solutions, which often defy the logic of alliances, and may even contradict alliance rationality. The calculus of Non-Traditional Security may overlap with the logic of arms control in that political and military hostility has less priority than the need for cooperative solutions.

Notes

1 Research for this chapter was funded in part by the Department of Foreign Affairs and International Trade (Canada). Views herein are those of the author and do not necessarily represent those of the Government of Canada.
2 The source of the estimate was the National Security Planning Agency, and without some independent verification, must be accepted with caution. Another report indicated 152 000 political prisoners. (*Yonhapp*, 1990 January 17, in *DR:EA*. 1990 January 17, 16_/ *A,mestu Omtermatopm* reported that there were over 100 000 political prisoners in North Korea. (*Chosen Ilbo, 1989* October 22, 2 in *DR:EA*, 1989 October 26, 27).
3 A *Far Eastern Economic Review* editorial traces the tendency back to the 1948 UN Declaration on Human Rights, 'which guarantees everything from "right" to social security to a right to a "free" and "compulsory" education'. (1993 June 17, 5).
4 Note that President Kim Young Sam attacked the system of past corruption by refusing to accept any political donations, and the Public Officials Ethics Law requires top officials to disclose their wealth (*Far Eastern Economic Review*, 1993 June 24, 18).
5 For a survey of North Korea's strategy and organization for anti-South terrorism, see the translation of an article by Captain Yi Tae-yun *Vision*, No. 9, October 1989, pp. 80–4, in *DR:EA*, 1990 January 3, 17–21).

8
Trends in Transboundary Water Disputes and Dispute Resolution

Aaron T. Wolf and Jesse H. Hamner

Background

'Water' and 'war' are two topics being assessed together with increasing frequency. The approximately 250 international watersheds cover more than one half of the land surface of the globe, and affect 40 per cent of its population. Water is a resource which ignores political boundaries, fluctuates in both space and time, has multiple and conflicting demands on its use, and whose international law is poorly developed, contradictory and unenforceable. As a consequence, recent articles in the academic literature (Cooley, 1984; Starr, 1991; Gleick, 1993; and others) and popular press (Bulloch and Darwish, 1993; World Press Review, 1995) point to water not only as a cause of historic armed conflict, but as *the* resource which will bring combatants to the battlefield in the twenty-first century.

The historic reality has been quite different from what the 'water wars' literature would have one believe. In modern history, only seven minor skirmishes have been waged over international waters – invariably other inter-related issues also factor in. Conversely, over 3600 treaties have been signed over different aspects of international waters, many showing tremendous elegance and creativity for dealing with this critical resource.

Despite the fact that countries seem not to go to war over water, the relationship between water scarcity and acute conflict still needs to be assessed. However, such an investigation should focus on more subtle relations between water and its users: historic evidence *does* suggest, for example, a relationship between access to clean water supplies and political stability. No better example exists, perhaps, than the Bangladeshi relationship with the Ganges: as the river's flow decreased as a consequence of Indian diversions, not only did severe environmental degradation result, but farmers and town-dwellers whose water supply became salinized formed a wave of environmental refugees, many thousands of whom poured into India. (Biswas and Hashimoto, 1996) This water-induced instability has recurred throughout the Middle East, Africa and Asia.

Once cooperative water regimes are established through treaty, however, they turn out to be tremendously resilient over time, even between otherwise hostile riparians, and even as conflict is waged over other issues. International water regimes have continued to function in the Mekong basin since 1957, despite the Vietnam War; the Jordan basin (between Israel and Jordan) since 1955, even as these riparians until only recently were in a legal state of war; and along the Indus, even through India–Pakistan warfare.

Given this critical role of transboundary water agreements, the processes which lead to their necessity, negotiation, and structure have been surprisingly poorly studied. This chapter describes three components of transboundary water issues. The first section investigates the history of transboundary water conflicts and suggests that the 'water wars' literature simply is not based on an historic reality. The second section offers insight into the much richer history of water dispute resolution as exemplified in water treaties which have been negotiated over time. The major findings of the Transboundary Freshwater Dispute Database are presented as evidence for the cooperation-inducing characteristics of transboundary waters. Nevertheless, argument for the future based on the past would be disingenuous. Moreover, security issues are much broader than 'simple' questions of war and peace. Water, while not leading to warfare, has tremendous impacts on regional security issues. In light of this, the third section describes possible indicators to anticipate future water-related tension, based on 14 detailed case studies of the Database.

'Water wars' and water reality[1]

As mentioned earlier, there is a growing literature which describes water both as an historic and, by extrapolation, as a future cause of interstate warfare. Westing (1986) suggests that, 'competition for limited . . . freshwater . . . leads to severe political tensions and even to war'; Gleick (1993) describes water resources as military and political goals, using the Jordan and Nile as examples; Remans (1995) uses case studies from the Middle East, South Asia, and South America as 'well-known examples' of water as a cause of armed conflict; Samson and Charrier (1997) write that, 'a number of conflicts linked to freshwater are already apparent', and suggest that, 'growing conflict looms ahead'; Butts (1997) suggests that, 'history is replete with examples of violent conflict over water', and names four Middle Eastern water sources particularly at risk; and Homer-Dixon (1994), citing the Jordan and other water disputes, comes to the conclusion that 'the renewable resource most likely to stimulate interstate resource war is river water'.

A close examination of the case studies cited as historic interstate water conflict suggests some looseness in classification. Samson and Charrier (1997), for example, list 18 cases of water disputes, only one of which is described as 'armed conflict', and that particular case (on the Cenepa River) turns out not to be about water at all but rather about the location of a shared boundary which happens to coincide with the watershed. Armed conflict did not take place in any of Remans's (1995) 'well-known' cases (save the one between Israel and Syria, described below), nor in any of the other lists of water-related tensions presented.

The examples most widely cited are wars between Israel and her neighbors. Westing (1986) lists the Jordan river as a cause of the 1967 war and, in the same volume, Falkenmark (1986), mostly citing Cooley (1984), describes water as a causal factor in both the 1967 war and the 1982 Israeli invasion of Lebanon. Myers (1993), citing Middle East water as his first example of 'ultimate security', writes that 'Israel started the 1967 war in part because the Arabs were planning to divert the waters of the Jordan River system.' In fact, in the years since Israel's invasion of Lebanon in 1982, a 'hydraulic imperative' theory, which describes the quest for water resources as *the* motivator for Israeli military conquests, both in Lebanon in 1979 and 1982 and earlier, on the Golan Heights and West Bank in 1967, was developed in the academic literature and the popular press (see, for example, Davis *et al.*, 1980; Stauffer, 1982; Schmida, 1983; Stork, 1983; Cooley, 1984; Dillman, 1989; Beaumont, 1991).

The only problem with these theories is a complete lack of evidence. While shots were fired over water between Israel and Syria from 1951–53 and 1964–66, the final exchange, including both tanks and aircraft on July 14, 1966, stopped Syrian construction of the diversion project in dispute, effectively ending water-related tensions between the two states – the 1967 war broke out almost a year later. The 1982 invasion provides even less evidence of any relation between hydrologic and military decision-making. In extensive papers investigating precisely such a linkage between hydro-strategic and geo-strategic considerations, both Libiszewski (1995) and Wolf (1995) conclude that water was neither a cause nor a goal of any Arab–Israeli warfare.

To be fair, I should note that I am only describing the relationship between interstate armed conflict and water as a scarce resource for human consumption. I exclude both intrastate disputes, as well as those where water was a means, method, or victim of warfare. I also exclude disputes where water is incidental to the dispute, such as those about fishing rights, access to ports, transportation, or river boundaries. Many of the authors I cite, notably Gleick (1993), Libiszewski (1995) and Remans (1995), are very careful about these distinctions. The bulk of the articles cited above, then, turn out to be about political tensions or stability rather than about warfare, or about water as a tool, target, or victim of armed conflict; while important issues, they are not the same as 'water wars'.

In order to cut through the prevailing anecdotal approach to the history of water conflicts, we investigated those cases of international conflict where armed exchange was threatened or took place over water resources *per se*. We utilized the most systematic collection of international conflict – the International Crisis Behavior (ICB) dataset (Brecher and Wilkenfeld, 1997), which contains only those disputes which were considered to be international crises by the principal investigators. Their definition of an international crisis is any dispute where (1) basic national values are threatened (eg territory, influence or existence); (2) time for making decisions is limited; and (3) the probability for military hostilities is high. Using these guidelines, they identified 412 crises for the period 1918–94. Joey Hewitt, of the University of Maryland at College Park, searched the text files of the ICB dataset for water-related key-words, and found four disputes where water was at least partially a cause. These have been researched and supplemented by three others at the University of Alabama. The complete list includes seven disputes:

1 1948 – Partition between India and Pakistan leaves the Indus basin divided in a particularly convoluted fashion. Disputes over irrigation water exacerbate tensions in the still-sensitive Kashmir region, bringing the two riparians 'to the brink of war'. Twelve years of World Bank – led negotiations lead to the 1960 Indus Waters Agreement.

2 February 1951 – September 1953. Syria and Israel exchange sporadic fire over Israeli water development works in the Huleh basin, which lies in the demilitarized zone between the two countries. Israel moves its water intake to the Sea of Galilee.
3 January – April 1958. Amidst pending negotiations over the Nile waters, Sudanese general elections, and an Egyptian vote on Sudan–Egypt unification, Egypt sends an unsuccessful military expedition into territory in dispute between the two countries. Tensions were eased (and a Nile Waters Treaty signed) when a pro-Egyptian government was elected in Sudan.
4 June 1963 – March 1964. 1948 boundaries left Somali nomads under Ethiopian rule. Border skirmishes between Somalia and Ethiopia are over disputed territory in Ogaden desert, which includes some critical water resources (both sides are also aware of oil resources in the region). Several hundred are killed before cease-fire is negotiated.
5 March 1965 – July 1966. Israel and Syria exchange fire over 'all-Arab' plan to divert the Jordan River headwaters, presumably to pre-empt Israeli 'national water carrier', an out-of-basin diversion plan from the Sea of Galilee. Construction of the Syrian diversion is halted in July 1966.
6 April – August 1975. In a particularly low-flow year along the Euphrates, as upstream dams are being filled, Iraqis claim that the flow reaching its territory was 'intolerable', and asked that the Arab League intervene. The Syrians claim that less than half the river's normal flow are reaching *its* borders that year and, after a barrage of mutually hostile statements, pull out of an Arab League technical committee formed to mediate the conflict. In May 1975, Syria closes its airspace to Iraqi flights and both Syria and Iraq reportedly transfer troops to their mutual border. Only mediation on the part of Saudi Arabia breaks the increasing tension.
7 April 1989 – July 1991. Two Senegalese peasants were killed over grazing rights along the Senegal River, which forms the boundary between Mauritania and Senegal, sparking smoldering ethnic and land reform tensions in the region. Several hundred are killed as civilians from border towns on either side of the river attack each other before each country uses its army to restore order. Sporadic violence breaks out until diplomatic relations are restored in 1991.

As we see, the actual history of armed water conflict is somewhat less dramatic then the 'water wars' literature would lead one to believe: a total of seven incidents, in three of which no shots were fired. As near as we can find, *there has never been a single war fought over water.*[2]

This is not to say there is no history of water-related violence – quite the opposite is true – only that these incidents are at the sub-national level, generally between tribes, water-use sectors, or states. Examples of internal water conflicts, in fact, are quite prevalent, from interstate violence and death along the Cauvery River in India, to California farmers blowing up a pipeline meant for Los Angeles, to much of the violent history in the Americas between indigenous peoples and European settlers. The desert state of Arizona even commissioned a navy (made up of one ferry boat) and sent its state militia to stop a dam and diversion on the Colorado River in 1934 (Fredkin, 1981).

In addition, one need look no further than relations between India and Bangladesh to note that internal instability can both be caused by, and exacerbate, international water disputes. At issue is a barrage which India has built at Farakka, which diverts a portion of the Ganges flow away from its course into Bangladesh, towards Calcutta 100 miles to the south, in order to flush silt away from that city's seaport. Adverse effects in Bangladesh resulting from reduced upstream flow have included degradation of both surface and groundwater, change in morphology, impeded navigation, increased salinity, degraded fisheries, and danger to water supplies and public health. Environmental refugees out of affected areas have further compounded the problem. Ironically, many of those displaced in Bangladesh have found refuge in India (Biswas and Hashimoto, 1996).

So, while no 'water wars' have occurred, there is ample evidence that the lack of clean freshwater has lead to occasionally intense political instability and that, on a small scale, acute violence can result. What we seem to be finding, in fact, is that geographic scale and intensity of conflict are *inversely* related.

The transboundary freshwater dispute database

The UN Food and Agriculture Organization has identified more than 3600 treaties relating to international water resources dating between AD 805 and 1984, the majority of which deal with some aspect of navigation (UN FAO, 1978, 1984). Since 1814, states have negotiated approximately 300 treaties which deal with non-navigational issues of water management, flood control, hydropower projects or allocations for consumptive or non-consumptive uses in international basins. Including only those signed in this century which deal with water *per se*, and excluding those which deal only with boundaries, navigation or fishing rights, the authors have collected full and partial texts of 145 treaties in a Transboundary Freshwater Dispute Database at the University of Alabama. The collection efforts continue in an ongoing project of the Department of Geography and the Center for Freshwater Studies, in conjunction with projects funded by the World Bank and the US Institute of Peace.

Negotiating notes and published descriptions of treaty negotiations are also being collected in the Database – in 1998 the Database contained 14 detailed case studies. These cases include nine watersheds – the Danube, Euphrates, Jordan, Ganges, Indus, Mekong, Nile, La Plata, and Salween; two sets of aquifer systems – US–Mexico shared systems and the West Bank Aquifers; two lake systems – the Aral Sea and the Great Lakes; and one engineering works – the Lesotho Highlands Project.

To date, few authors have undertaken systematic work on the body of international water treaties as a whole, although some use treaty examples to make points about specific conflicts, areas of cooperation, or larger issues of water law (see, for example, Vlachos, 1990; Teclaff, 1991; McCaffrey, 1993; Eaton and Eaton, 1994; Housen-Couriel, 1994; Dellapenna, 1995; Kliot *et al.*, 1997). In two of the most thorough exceptions, Dellapenna (1994) describes the evolution of treaty practice dating back to the mid-1800's, and Wescoat (1995) assesses historic trends of water treaties dating from 1648–1948 in a global perspective. In addition, McCaffrey (1993) offers theories about trends in treaty-making, specifically the move towards integrated management from unilateral development, the move away from navigation as the primary use, and the trend towards 'equitable utilization'. Hayton (1988, 1991) has argued that international law should account more for hydrologic processes.

Treaties can tell about regional hegemony, about how and which water needs are met, about the relative importance of water in the political climate, about development issues, and whether earlier treaties have successfully guided or guaranteed state behavior. To organize and analyse these treaties, Jesse Hamner has developed a systematic computer compilation which catalogs the treaties by basin, countries involved, date signed, treaty topic, allocations measure, conflict resolution mechanisms, and non-water linkages.

Major findings

The contents of the Database are qualitatively and quantitatively assessed for their provisions regarding the following criteria: basin involved; principal focus; number of signatories; non-water linkages (such as money, land or concessions in exchange for water supply or access to water); provisions for monitoring, enforcement and conflict resolution; method and amount of water division, if any; and the date signed (see Table 8.1 for treaty dates and titles) Preliminary descriptions of our findings follow (see Table 8.2 for a statistical summary)

Treaty signatories

One hundred and twenty-four of the 145 treaties (86 per cent) are bilateral. Twenty-one (14 per cent) are multilateral; two of the multilateral treaties are unsigned agreements or drafts.

It is unclear whether so many treaties are bilateral because only two states share a majority of international watersheds or because, according to negotiation theory, the difficulty of negotiations increases as the number of parties increases (Zartman, 1991, 65–77).

In multinational basins, this preference towards bilateral agreements can preclude the regional management long advocated by water resource managers. One who ignores the watershed as the fundamental planning unit, where surface- and groundwater, quality and quantity are all interrelated, also ignores hydrologic reality. The Jordan basin, for example, has been characterized by bilateral arrangements – the only regional talks on the basin, the Johnston negotiations of 1953–55, went unratified. As unilateral development in the basin proceeded in the absence of agreement, each state's goals and plans abutted against those of the other co-riparians, leading to inefficient development and even to exchanges of fire in the early 1950's and mid-1960's. Similarly, India has a long-standing policy of adhering to bilateral negotiations, presumably because it can best address its own needs *vis-à-vis* each of its neighbors separately. Partly as a consequence, neither the Ganges–Brahmaputra nor the Indus River systems have ever been managed to their potential efficiency. All but three multilateral agreements lack definite allotments, although a few establish advisory and governing bodies among states.

Of the 21 multilateral treaties/agreements, developing nations account for 13. Only one multilateral treaty that regarding water withdrawals from Lake Constance signed by Germany, Austria, and

Table 8.1 Shortened treaty titles, listed chronologically

Date	Title
7/20/1874	Articles of agreement between the Edur Durbar and the British government
2/26/1885	Act of Berlin
8/10/1889	Agreement between Great Britain and France
4/15/1891	Protocol between Great Britain and Italy for the Demarcation of their respective spheres
9/16/1892	Amended Terms of Agreement between the British Government and the State of Jind, for regulating the supply of water for irrigation
8/29/1893	Agreement between the British government and the Patiala state regarding the Sirsa branch of the Western Jumna canal
2/4/1895	Exchange of letters between Great Britain and France
3/18/1902	Exchange of notes between Great Britain and Ethiopia
2/23/1904	Final working agreement relative to the Sirhind canal between Great Britain and Patiala, Jind and Nabha
5/9/1906	Agreement... modifying the Agreement signed at Brussels 12 May 1894
10/19/1906	Agreement between Great Britain and France
4/11/1910	Convention regarding the Water Supply of Aden between Great Britain and the Sultan of Abdali
5/5/1910	Treaty Between Great Britain and the United States relating to Boundary Waters and Boundary Questions
9/4/1913	Exchange of notes constituting an agreement... respecting the boundary between Sierra Leone and French Guinea
6/12/1915	Protocol... for the delimitation of the frontier along the River Horgos
4/20/1921	Convention of Barcelona
10/28/1922	Convention between [Finland] and the [USSR] concerning the maintenance of river channels and the regulation of fishing on water courses
2/14/1925	Convention between [Norway] and [Finland] concerning the international legal regime of the waters of the Pasvik (Paatsjoki) and the Jakobselv
2/24/1925	Agreement between the United States of America and Canada to regulate the level of Lake of the Woods
6/15/1925	Notes exchanged... respecting the regulation of the utilisation of the waters of the river Gash
12/20/1925	Exchange of Notes between Great Britain and Italy
7/1/1926	Agreement... regulating the use of the water of the Cunene River
7/20/1927	Convention... regarding various questions of economic interest
8/11/1927	Convention between Spain and Portugal to regulate the hydro-electric development of the international section of the River Douro

Table 8.1 *continued*

Date	Description
1/29/1928	Convention between the German Reich and the Lithuanian Republic regarding the maintenance and administration of the frontier waterways
5/7/1929	Exchange of Notes... in regard to the use of... the river Nile for irrigation purposes
4/29/1931	Exchange of Notes... respecting the boundary between the mandated territory of South Africa and Angola
11/22/1934	Agreement... regarding water rights on the boundary between Tanganyika and Ruanda–Urundi
5/11/1936	Exchange of notes... regarding the boundary between Tanganyika Territory and Mozambique
11/7/1940	Exchange of notes between... the United States of America and... Canada constituting an Agreement regarding the development of certain portions...
5/20/1941	Exchange of Notes between the government of the United States and the Government of Canada... concerning temporary diversion for power
11/27/1941	Exchange of Notes constituting an agreement between the government of the United States and the Government of Canada relating to additional...
5/22/1944	Declaration and Exchange of Notes concerning the Termination of the Process of Demarcation of the Peruvian–Ecuadorean Frontier
11/14/1944	Treaty Between the United States of America and Mexico Relating to the Waters of the Colorado and Tijuana Rivers, and of the Rio Grande
6/1/1945	Supplementary boundary treaty between [Argentina] and [Paraguay] on the river Pilcomayo
12/30/1946	Agreement concerning the utilization of the rapids of the Uruguay River in the Salto Grande area
2/3/1947	Treaty between the [USSR] and [Finland] on the transfer to the territory of the Soviet Union of part of the state territory of Finland in the region of
2/10/1947	Treaty of Peace with Italy, Signed at Paris, on 10 February 1947
5/4/1948	Inter-Dominion Agreement Between the Government of India and the Government of Pakistan, on the Canal Water Dispute Between
5/31/1949	Exchanges of Notes... regarding the construction of the Owen falls Dam, Uganda
11/25/1949	Treaty concerning the Regime of the Soviet-Romanian State Frontier and Final Protocol
12/5/1949	Exchange of Notes Constituting an Agreement Between [Great Britain]... and [Egypt] regarding the construction of the Owen Falls Dam, Uganda
1/19/1950	Exchange of notes constituting an agreement between [Great Britain] (on behalf of... Uganda) and [Egypt] regarding cooperation in meteorological...

Table 8.1 *continued*

Date	Description
2/24/1950	Treaty between the [USSR] and [Hungary] concerning the regime of the Soviet-Hungarian state frontier and final protocol
2/27/1950	Treaty between the United States of America and Canada relating to the uses of the waters of the Niagara River
4/25/1950	State Treaty concerning the construction of a hydro-electric power-plant on the Sauer at Rosport/Ralingen
6/9/1950	Convention between the [USSR] and [Hungary] concerning measures to prevent floods and to regulate the water regime in the area of the frontier
9/7/1950	Terms of reference of the Helmand River Delta Commission and an interpretive statement relative thereto, agreed by conferees of...
10/16/1950	Agreement concerning the diversion of water in the Rissbach, Durrach and Walchen Districts
10/16/1950	Agreement between [Austria] and [Germany] concerning the Österreichisch-Bayerische Kraftwerke AG
4/18/1951	Letters between the irrigation adviser and director of irrigation, Sudan Government, and the controller of agriculture, Eritrea
4/25/1951	Agreement between [Finland] and [Norway] on the transfer from the course of the Näätämo (Neiden) river to the course of the Gandvik River
2/13/1952	Agreement concerning the Donaukraftwerk-Jochenstein Aktiengesellschaft
6/30/1952	Exchange of notes constituting an agreement between Canada and the United States of America relating to the St. Lawrence Seaway Project
7/16/1952	Exchange of notes constituting an agreement between the [UK/Uganda] and [Egypt] regarding the construction of the Owen Falls Dam in Uganda
12/25/1952	Convention between the [USSR] and [Romania] concerning measures to prevent floods and to regulate the water regime of the river Prut
1/21/1953	Exchange of notes constituting an agreement between [Great Britain] and [Portugal] providing for the Portuguese participation in the Shiré valley
6/4/1953	Agreement between the Republic of Syria and the Hashemite Kingdom of Jordan concerning the utilization of the Yarmuk waters.
11/12/1953	Exchange of notes constituting an agreement between the United States and Canada relating to the establishment of the St Lawrence River joint
4/16/1954	Agreement between [Czechoslovakia] and [Hungary] concerning the settlement of technical and economic questions relating to frontier water
4/25/1954	Agreement between the Government of India and the Government of Nepal on the Kosi Project

Table 8.1 *continued*

Date	Description
5/25/1954	Convention between the Governments of [Yugoslavia] and [Austria] concerning water economy questions relating to the Drava
11/18/1954	Agreement between [Great Britain / Rhodesia-Nyasaland] with regard to certain... natives living on the Kwando River
12/16/1954	Agreement between [Yugoslavia] and [Austria] concerning water economy questions in respect of the frontier sector of the Mura
4/7/1955	Agreement between [Yugoslavia] and [Romania] concerning questions of water control on water control systems and watercourses on or intersected...
4/20/1955	Exchange of notes between Peru and Bolivia establishing a joint commission for study of... joint use of the waters of Lake Titicaca
8/8/1955	Agreement between [Yugoslavia] and [Hungary] together with the statute of the Yugoslav-Hungarian water economy commission
12/31/1955	Johnston Negotiations
1/20/1956	Agreement concerning cooperation between [Brazil] and [Paraguay] in a study on the utilization of the water power of the Acaray and Monday
4/9/1956	Treaty between the Hungarian People's Republic and the Republic of Austria concerning the regulation of water economy questions
8/18/1956	Agreement between the [USSR] and [China] on joint research operations to determine the natural resources of the Amur river basin and the prospects
10/13/1956	Treaty between [Czechoslovakia] and [Hungary] concerning the regime of state frontiers.
12/5/1956	Agreement between [Yugoslavia] and [Albania] concerning water economy questions, together with the statute of the Yugoslav–Albanian Water
2/19/1957	Agreement between Bolivia and Peru concerning a preliminary economic study of the joint utilization of the waters of Lake Titicaca.
5/14/1957	Treaty between the Government of the [USSR] and [Iran] concerning the regime of the Soviet-Iranian Frontier and the procedure for the settlement
8/11/1957	Agreement between Iran and the Soviet Union for the joint utilisation of the frontier parts of the rivers Aras and Atrak for irrigation and power
12/18/1957	Agreement between Norway and the Union of Soviet Socialist Republics on the Utilization of Water power on the Pasvik (Paatso) River
1/23/1958	Agreement between [Argentina] and [Paraguay] concerning a study of the utilization of the water power of the Apipe Falls

Table 8.1 *continued*

Date	Description
3/21/1958	Agreement between the [Czechoslovakia] and [Poland] concerning the use of water resources in frontier waters
4/4/1958	Agreement concerning water-economy questions between the government of [Yugoslavia] and [Bulgaria]
7/10/1958	State treaty between [Luxembourg] and [West Germany] concerning the construction of hydroelectric power-installations on the Our River
7/12/1958	Agreement between the Government of the French Republic and the Spanish Government relating to Lake Lanoux
4/29/1959	Agreement between the [USSR], [Norway], and [Finland] concerning the regulation of Lake Inari by means of the Kaiakoski... dam
10/23/1959	Indo-Pakistan agreement (with appendices) on East Pakistan border disputes
11/8/1959	Agreement between the Government of the United Arab Republic and the government of Sudan
12/4/1959	Agreement Between [Nepal] and [India] on the Gandak Irrigation and Power Project
1/11/1960	Agreement between Pakistan and India on West Pakistan–India Border Disputes
9/19/1960	Indus Waters Treaty
10/24/1960	Agreement relating to the construction of Amistad Dam on the Rio Grande to form part of the system of international storage dams provided for by the...
1/17/1961	Treaty relating to cooperative development of the water resources of the Columbia River Basin (with annexes)
02/24/1961	Exchange of notes constituting an agreement concerning the treaty of 12 May 1863 to regulate the diversion of water from the River Meuse and the...
4/26/1963	Exchange of notes constituting an agreement... for the development of the Mirim Lagoon
7/26/1963	Convention of Bamako
10/26/1963	Act... States of the Niger Basin
11/25/1963	Agreement... relating to the Central African Power Corporation
11/30/1963	Convention between [Yugoslavia] and [Romania] concerning the operation of the Iron Gates water power and navigation
11/30/1963	Agreement between the Socialist Federal Republic of Yugoslavia and the Romanian People's Republic concerning the construction and operation...
11/30/1963	Convention between the Socialist Federal Republic of Yugoslavia and the Romanian People's Republic concerning compensation for damage
1/22/1964	Exchange of notes constituting an agreement between Canada and the United States of America concerning the treaty relating to cooperative...

Table 8.1 continued

Date	Description
1/22/1964	Exchange of notes constituting an agreement between Canada and the United States of America regarding sale of Canada's entitlement
2/11/1964	Agreement between Iraq and Kuwait concerning the supply of Kuwait with fresh water
5/22/1964	Convention and Statutes... Lake Chad Basin
7/16/1964	Convenio entre España y Portugal para Regular el Aprovechamiento hydroelectrico de los tramos internacionales de rio Duero y de sus afluentes
7/17/1964	Agreement between [Poland] and the [USSR] concerning the use of water resources in frontier waters
9/16/1964	Exchange of notes constituting an agreement between Canada and the United States of America authorizing the Canadian entitlement purchase
11/25/1964	Agreement concerning the river Niger commission and the navigation and transport on the river Niger
8/12/1965	Convention between Laos and Thailand for the supply of power
4/30/1966	Agreement between [West Germany], [Austria], and [Switzerland] relating to the withdrawal of water from Lake Constance
8/24/1966	Exchange of notes constituting an agreement concerning the loan of waters of the Colorado River for irrigation of lands in the Mexicali Valley.
12/19/1966	Revised Agreement Between [Nepal] and [India] on the Kosi Project
4/1/1967	Untitled: Agreement between South Africa and Portugal
9/28/1967	Franco-Italian convention concerning the supply of water to the Commune of Menton
12/7/1967	Treaty between [Austria] and [Czechoslovakia] concerning the regulation of water management questions relating to frontier waters
2/27/1968	Agreement between [Czechoslovakia] and [Hungary] concerning the establishment of a river administration in the Rajka-Gönyü Sector
5/29/1968	Convenio y Protocola Adicional Para Regular el Uso y aprovechamiento hidraulico de los tramos internacionales de los rios Miño, Limia, Tajo
10/23/1968	Agreement between the People's Republic of Bulgaria and the Republic of Turkey concerning cooperation
1/21/1969	Agreement between South Africa and Portugal
3/21/1969	Exchange of notes constituting an agreement for the construction of a temporary cofferdam at Niagara
3/21/1969	Exchange of notes constituting an agreement between Canada and the United States of America for the temporary diversion for power purposes
7/4/1969	Convention concerning development of the Rhine between Strasbourg and Lauterbourg

Table 8.1 *continued*

Date	Agreement
1/30/1970	Convention of Dakar
12/16/1971	Agreement Between [Romania] and the [USSR] on the joint construction of the Stinca-Costesti Hydraulic Engineering Scheme
7/12/1972	Agreement between [Finland] and the [USSR] concerning the production of electric power in the part of the Vuoksi river bounded by the Imatra
11/24/1972	Statute of the Indo-Bangladesh Joint Rivers Commission
4/26/1973	Treaty between [Brazil] and [Paraguay] concerning the hydroelectric utilization of the water resources of the Parana river
11/13/1973	Agreement between [Australia / Papua New Guinea] and [Indonesia] concerning administrative border arrangements
1/31/1975	Joint declaration of principles for utilization of the waters of the lower Mekong basin, signed by [Cambodia], [Laos], [Thailand], and [Vietnam]
3/6/1975	Agreement . . . concerning the use of frontier watercourses
2/12/1976	Segundo protocolo
11/5/1977	Agreement between [Bangladesh] and [India] on sharing of the Ganges' waters at Farakka and on augmenting its flows
4/7/1978	Agreement between [Nepal] and [India] on the Renovation and Extension of Chandra Canal, Pumped Canal, and Distribution of the Western Kosi Canal
6/30/1978	Convention relating to the creation of the Gambia River Basin Development Organization
7/3/1978	Treaty for Amazonian Cooperation
10/19/1979	Agreement on Paraná River projects
11/21/1980	Convention creating the Niger Basin Authority
7/20/1983	Meeting of the Joint Rivers Commission
10/01/1986	Treaty on the Lesotho Highlands Water Project between [Lesotho] and [South Africa]
10/8/1990	Convention . . . on the international commission for the protection of the Elbe
3/26/1993	Agreement on joint activities in addressing the Aral Sea
06/30/1994	Draft Convention on Cooperation for the Protection and Sustainable Use of the Danube River
10/26/1994	Treaty of peace between [Israel] and [Jordan], done at Arava/Araba crossing point
3/3/1995	Resolution of the Heads of States of the Central Asia [*sic*] on work of the EC of ICAS on implementation
04/05/1995	Agreement on the Cooperation for the Sustainable Development of the Mekong River Basin
9/28/1995	Israeli-Palestinian Interim Agreement on the West Bank and the Gaza Strip
12/12/1996	Treaty between [India] and [Bangladesh] on Sharing of the Ganga/Ganges Waters at Farakka

Table 8.2 Treaty statistics summary sheet

Signatories	Enforcement
Bilateral 124/145 (86%)	Council 26/145 (18%)
Multilateral 21/145 (14%)	None/Not available 116/145 (80%)
Principle Focus	Force 2/145 (1%)
Water Supply 53/145 (37%)	Economic 1/145 (<1%)
Hydropower 57/145 (39%)	**Unequal power relationship**
Flood control 13/145 (9%)	Yes 52/145 (36%)
Industrial uses 9/145 (6%)	No/Unclear 93/145 (64%)
Navigation 6/145 (4%)	**Information sharing**
Pollution 6/145 (4%)	Yes 93/145 (64%)
Fishing 1/145 (<1%)	No/Not available 52/145 (36%)
Monitoring	**Water allocation**
Provided 78/145 (54%)	Equal portions 15/145 (10%)
No/Not Available 67/145 (46%)	Complex but clear 39/145 (27%)
Conflict resolution	Unclear 14/145 (10%)
Council 43/145 (30%)	None/Not available 77/145 (53%)
None/Not Available 79/145 (54%)	**Non-water linkages**
Other governmental unit 9/145 (6%)	Money 44/145 (30%)
	Land 6/145 (4%)
United Nations/Third party 14/145 (10%)	Political concessions 2/145 (1%)
	Other linkages 10/145 (7%)
	No linkages 83/145 (57%)

Switzerland in 1966 exists among industrialized nations for access to a water source. None of the preindustrial-nation multilateral agreements specified any water allocations – all involved hydropower and/or industrial uses.

The states surrounding the Aral Sea have an agreement, dated 1993, that deals with several joint issues, but the text lacks allocations, and provides too little detail for planning water use. Like the Aral Sea, Lake Chad also suffers from poorly managed use and current deficit water withdrawals. The Chad Basin treaty (1964), among Cameroon, Niger, Nigeria and Chad, deals with economic development inside the basin, the lake's tributaries, and industrial uses of the lake, but lacks allocations. The agreement does create a commission, which, among other things, arbitrates disputes concerning implementation of the treaty. The commission prepares general regulations, coordinates the research activities of the four states, examines their development schemes, makes recommendations, and maintains contact among the four states.

Principal focus

Most treaties focus on hydropower and water supplies: 57 (39 per cent) of the treaties discuss hydroelectric generation, and 53 (37 per cent) distribute water supplies. Nine (6 per cent) mention industrial uses, six (4 per cent) navigation, and six (4 per cent) primarily discuss pollution. Thirteen of the 145 (9 per cent) focus on flood control. The Database includes one treaty which discusses fishing primarily (<1 per cent).

Great Britain is a signatory in 27 (19 per cent) of the treaties. Of that number, 17 (63 per cent) relate to water supply, all but one (in Canada) of an African or South Asian colony.

Monitoring

Seventy-eight treaties (54 per cent) have provisions for monitoring, while 67 (46 per cent) do not. When monitoring is mentioned, it is addressed in detail, often including provisions for data-sharing, surveying, and schedules for collecting data.

Information-sharing generally engenders good will and can provide confidence-building measures between co-riparians. Unfortunately, some states classify river flows as secrets, and others use the lack of mutually agreeable data as a stalling technique in their negotiations (Kaye, 1989, 17). Most monitoring clauses contain only the most rudimentary elements, perhaps due to the time and labor costs of gathering data.

However, data collected by signatories of the treaty can provide a solid base for later discussions. India and Bangladesh eventually agreed on Ganges flow data and based a workable agreement on that data in 1977, where previously both parties could not agree to the accuracy of each other's hydrologic records. The cooperation between engineers or among council members can result in the formation of an epistemic community – a community of trans-boundary professionals – another positive outcome of data gathering/sharing. Treaties have not commonly included provisions to monitor compliance, but such additions may bolster trust and increase the strength of these epistemic bonds.

Method for water division

Few treaties allocate water: clearly defined allocations account for 54 (37 per cent) of the agreements. Of that number, 15 (28 per cent) specify equal portions, and 39 (72 per cent) provide a specific means of allocations. In other work, Wolf (1999a) finds four general trends in those treaties which specify allocations: (1) a shift in positions often occurs during negotiations from 'rights-based' criteria, whether hydrography or chronology, in favor of 'needs-based' values, based on irrigable land or population, for example; (2) In disputes between upstream and downstream riparians concerning existing and future uses, the needs of the down-stream riparian are more-often delineated – upstream needs are mentioned only in boundary waters accords in humid regions – and existing uses, when mentioned, are *always* protected; (3) Economic benefits have not been explicitly used in allocating water, although economic principles have helped guide definitions of 'beneficial' uses and have suggested 'baskets' of benefits, including both water and non-water resources, for positive-sum solutions; (4) The uniqueness of each basin is repeatedly suggested, both implicitly and explicitly, in the treaty texts.

This last point is exemplified in the unique measures which negotiators have devised: the 1959 Nile Waters Treaty divides the average flow based on existing uses, then evenly divides any future supplies (projected from the Aswan High Dam and the Jonglei Canal Project); the Johnston negotiations led to allocations between Jordan riparians based on the irrigable land within the watershed – each party could then do what it wished with its allocations, including divert it out-of-basin; and the Boundary Waters Agreement, negotiated with a hydropower focus between Canada and the US, which allows a greater minimum flow of the Niagra River over the famous falls during summer daylight hours, when tourism is at its peak.

Hydropower

Fifty-seven of the treaties (39 per cent) focus on hydropower. Power-generating facilities bring development, and hydropower provides a cheap source of electricity to spur developing economies. Postel (1997) and others, however, suggest that the age of building dams will soon end because of lack of funding for large dams, a general lack of suitable new dam sites, and environmental concerns.

Not surprisingly, less-developed mountain-nations at the headwaters of the world's great rivers are signatories to the bulk of the hydropower agreements: Nepal alone, with an estimated 2 per cent of the world's hydropower potential (Aryal, 1995, 160), has four treaties with India (the Kosi River agreements, 1954, 1966, 1978, and the Gandak Power Project, 1959) to exploit the huge power potential in the region.

Groundwater

Only three agreements deal with groundwater supply: the 1910 convention between Great Britain and the Sultan of Abdali, the 1994 Jordan–Israel peace treaty, and the Palestinian–Israeli accords (Oslo II). Treaties that focus on pollution usually mention groundwater, but do not address the issue quantitatively.

The complexities of groundwater law have been described by more than a few authors (Hayton, 1982; Utton, 1982). Overpumping can destroy cropland through salinity problems, either by seawater intrusion or evaporation-deposition; therefore, allocating too much water (or overpumping) can decimate future freshwater supplies.

The Bellagio Draft Treaty, developed in 1989, attempts to provide a legal framework for groundwater negotiations. The treaty describes principles based on mutual respect, good neighborliness, and reciprocity, which requires joint management of shared aquifers (Hayton and Utton, 1989). While the Draft recognizes that obtaining groundwater data can prove difficult and expensive, and its acceptance relies on cooperative and reciprocal negotiations, it does provide a useful framework for future groundwater diplomacy.

Non-water linkages

Negotiators may facilitate success by enlarging the scope of water disputes to include non-water issues. If pollution causes trouble in a downstream country, an upstream neighbor may opt to pay for a treatment plant in lieu of reduced inputs or reduced withdrawals. In such a case, lesser amounts of high-quality water may improve relations more than a greater quantity of polluted or marginal-quality water. Such tactics 'enlarge the pie' of available water and other resources in a basin. Non-Water Linkages include capital, 44/145 (30 per cent); land, 6/145 (4 per cent); political concessions, 2 (1 per cent); other linkages, 10/145 (7 per cent); and no linkages, 83/145 (57 per cent).

In the 1929 Nile agreement, the British agreed to give technical support to both Sudan and Egypt. In lieu of payments, the Soviet Union agreed to compensate lost power generation to Finland in perpetuity (the 1972 Vuoksa agreement). Britain even established ferry service across newly-widened parts of the Hathmatee in India, in compensation for the inaccessibility problems created by its dam project in the late 1800s.

Compensation for land flooded by dam projects is common. British colonies usually agreed to pay for water delivery and reservoir upkeep. However, capital can provide compensation for a greater array of possibilities, flooded homes, and construction of new water facilities.[3]

Treaties which allocate water also include payments for water – 44 treaties (30 per cent) include monetary transfers or future payments. As early as 1925, Britain moved towards equitable use of the rivers in its colonies – Sudan agreed to pay a portion of the income generated by new irrigation projects to Eritrea, since the Gash river flowed through that state as well. Treaties also recognize the need to compensate for hydropower losses and irrigation losses due to reservoir storage.[4] Again, this fact emphasizes the monetary aspect of water: it does not describe water as a right or commodity.

Enforcement

Treaties handle disputes with technical commissions, basin commissions, or government officials. Fifty-two of the treaties (36 per cent) provide for an advisory council or conflict-addressing body within the government. Fourteen (10 per cent) refer disputes to a third-party or the United Nations. Thirty-two (22 per cent) make no provisions for dispute resolution, and 47 (32 per cent) of the texts are either incomplete or uncertain in the creation of dispute resolution mechanisms.

Historically, force or the threat of force can ensure that a water treaty will be followed[5] – power is an unfortunate guarantor of compliance. Britain could oversee its colonial water treaties because it had one of the most powerful administrative and military organizations in the world. Agreements on the Nile generally favor Egypt, while those on the Jordan River favor Israel for similar reasons.

While the conflict resolution mechanisms in these treaties do not generally show tremendous sophistication, new enforcement possibilities exist with new monitoring technology: it is now possible to manage a watershed in real time, using a combination of remote sensing and radio-operated control structures. In fact, the next major step in treaty development may well be mutually enforceable provisions, based in part on this technology.

Indicators of future water-related tensions

One can assume that each water treaty represents a dispute which has been resolved: some issue must arise for the parties to enter negotiations in the first place. It would stand to reason, then, that by assessing the variables which immediately preceded the negotiation of a treaty, one might be able to determine what factors act as indicators of impending dispute.

For example, many of the most virulent examples of water conflict came about immediately following the internationalization of a previously national waterway – such was the case on the Jordan, Indus, Nile and Aral basins. The existence of ethnic minorities or political sub-groups along major waterways, then, may point to regions with the potential for future hydropolitical stresses – Kurdish regions along the Euphrates or the Punjab between India and Pakistan, for example.

In contrast, and somewhat counter-intuitively, climate seems *not* to be a major variable in water disputes. This may be because water has multiple uses which are all critically important, but which change depending on climate conditions. The hydropower or transportation offered by a river in a humid climate, for example, is no less important to its riparians than is the irrigation water provided by a river in an arid zone. We see, then, hydropolitical disputes across the climatic spectrum, from the Plata and Ganges in the humid zone, to the Nile, Jordan, and Tigris–Euphrates, each of which have humid-zone headwaters, but which flow primarily through arid areas.

To do an extensive analysis of each of the basins represented by the treaties in our Database is beyond the scope of this paper, but the fourteen case studies mentioned above do suggest patterns which may be useful in allowing for anticipation of likely conflict.

In general, a pattern emerges as follows: riparians of an international basin implement water development projects unilaterally, first on water within their territory, in attempts to avoid the political intricacies of the shared resource. At some point, as water demand approaches supply, one of the riparians, generally the regional power,[6] will implement a project which impacts at least one of its neighbors. This might be to continue to meet existing uses in the face of decreasing relative water availability, as for example Egypt's plans for a high dam on the Nile, or Indian diversions of the Ganges to protect the port of Calcutta, or to meet new needs reflecting new agricultural policy, such as Turkey's GAP project on the Euphrates.

In the absence of relations or institutions conducive to conflict resolution, a project which impacts one's neighbors can become a flashpoint, as described above. Each of these projects is preceded by indicators of impending or likely water conflict, which might include:

Water quantity issues. Often, simply extrapolating water supply and demand curves will give an indication of when a conflict may occur, as the two curves approach each other. The mid-1960s, a period of water conflict in the Jordan basin, saw demand approaching supply in both Israel and Jordan. Also, major shifts in supply might indicate likely conflict, due to greater upstream use or, in the longer range, to global change. The former is currently the case both on the Mekong and on the Ganges. Likewise, shifts in demand, due to new agricultural policies or movements of refugees or immigrants can indicate problems. Water systems with a high degree of natural fluctuation can cause greater problems than relatively predictable systems.

Water quality issues. Any new source of pollution, or any new extensive agricultural development resulting in saline return flow to the system, can indicate water conflict. Arizona return flow into the Colorado was the issue over which Mexico sought to sue the US in the 1960s through the International Court of Justice, and is currently a point of contention on the lower Jordan between Israel, Jordan, and West Bank Palestinians.

Management for multiple use. Water is managed for a particular use or a combination of uses. A dam might be managed for storage of irrigation water, power generation, recreation or a combination. When the needs of riparians conflict, disputes are likely. Many upstream riparians, for instance, would manage the river within their territory primarily for hydropower where the primary needs of their downstream neighbors might be timely irrigation flows. Chinese plans for hydropower generation and/or Thai plans for irrigation diversions would have an impact on Vietnamese needs for both irrigation and better drainage in the Mekong Delta.

Political divisions. A common indicator of water conflict is shifting political divisions which reflect new riparian relations. Such is currently the case throughout Central Europe as national water bodies such as the Aral Sea, the Amu Dar'ya, and the Syr Dar'ya become international. Many of the conflicts presented here, including those on the Ganges, the Indus, and the Nile, took on international complications as the central authority of a hegemon, in these cases the British empire, dissipated.

Along with clues useful in anticipating whether or not water conflicts might occur, patterns based on past disputes may provide lessons for determining both the type and intensity of impending conflicts. These indicators might include:

Geopolitical setting. As mentioned above, relative power relationships, including riparian position, determine how a conflict unfolds (Lowi, 1993, 1995). A regional power which also has an upstream riparian position is in a better situation to implement projects which may become flashpoints for regional conflict. Turkey and India have been in such positions on the Euphrates and the Ganges, respectively. In contrast, the development plans of an upstream riparian may be held in check by a more powerful downstream riparian as have Ethiopia's plans for Nile development by Egypt.

The perception of unresolved issues with one's neighbors, both water-related and otherwise, may also be an exacerbating factor in water conflicts. Israel, Syria and Turkey, each and respectively have thorny political issues outstanding, which makes discussions on the Jordan and Euphrates more difficult.

Level of national development. Relative development can inform the nature of water disputes in a number of ways. For example, a more-developed region may have better options to alternative sources of water, or to different water management schemes, than less-developed regions, resulting in more options once negotiations begin. In the Middle East multilateral working group on water, for instance, a variety of technical and management options, such as desalination, drip irrigation, and moving water from agriculture to industry, have all been presented, which in turn supplement discussions over allocations of international water resources.

Different levels of development within a watershed, however, can exacerbate the hydropolitical setting. As a country develops, personal and industrial water demand tends to rise, as does demand for previously marginal agricultural areas. While this can be somewhat balanced by more access to water-saving technology, a developing country often will be the first to develop an international resource to meet its growing needs. Thailand has been making these needs clear with its relatively greater emphasis on Mekong development.

The hydropolitical issue at stake. In a survey of fourteen river basin conflicts, Mandel (1992) offers interesting insights relating the issue at stake with the intensity of a water conflict. He suggests that issues which include a border dispute in conjunction with a water dispute, such as the Shatt al-Arab waterway between Iran and Iraq and the Rio Grande between the US and Mexico, can induce more severe conflicts than issues of water quality, such as the Colorado, Danube and La Plata rivers. Likewise, conflicts triggered by human-initiated technological disruptions – dams and diversions – such as the Euphrates, Ganges, Indus and Nile, are more severe than those triggered by natural flooding, such as the Columbia and Senegal rivers.

One interesting lack of correlation is also found in Mandel's study – that between the number of disputants and the intensity of conflict. He suggests that this challenges the common notion that the more limited, in terms of number of parties involved, river disputes are easier to resolve.

Institutional control of water resources. An important aspect of international water conflicts is how water is controlled *within* each of the countries involved. Whether control of the resource is vested at the national level, as in the Middle East, the state level, as in India, or at the sub-state level, as in the United States, informs the complication of international dialog. Also, *where* control is vested institutionally is important. In Israel, for example, the Water Commissioner is under the authority of the Ministry of Agriculture, whereas Jordanian control is at the ministerial level, with the Ministry of Water. Theses respective institutional settings can make internal political dynamics quite different for similar issues.

National water ethos. This term incorporates several somewhat ambiguous parameters together which determine how a nation 'feels' about its water resources, which in turn can help determine how much it 'cares' about a water conflict. Some factors of a water ethos might include:

- 'mythology' of water in national history, eg Has water been the 'lifeblood of the nation?' Was the country built up around the heroic *fellah*? Is 'making the desert bloom' a national aspiration? In most countries, in contrast, water plays little role in the national history.
- importance of water/food security in political rhetoric;
- relative importance of agriculture versus industry in the national economy.

Conclusions

There is a large and growing literature warning of future 'water wars' – they point to water not only as a cause of historic armed conflict, but as *the* resource which will bring combatants to the battlefield in the twenty first century.

The historic reality has been quite different. In modern times, only seven minor skirmishes have been waged over international waters – invariably other interrelated issues also factor in. Conversely, over 3600 treaties have been signed over different aspects of international waters – 145 in this century alone on water *qua* water – many showing tremendous elegance and creativity for dealing with this critical resource. This is not to say that armed conflict has not taken place over water, only that such disputes generally are between tribe, water-use sector, or sub-national states. What we seem to be finding, in fact, is that geographic scale and intensity of conflict are *inversely* related.

War over water is neither strategically rational, hydrographically effective, nor economically viable. Shared interests along a waterway seem to overwhelm water's conflict-inducing characteristics and, once water management institutions are in place, they tend to be tremendously resilient. The patterns described in this chapter suggest that the more valuable lesson of international water is as a resource whose characteristics tend to induce cooperation, and incite violence only in the exception.

Nevertheless, 145 treaties which govern the world's international watersheds, and the international law on which they are based, are in their respective infancies. More than half of these treaties include no monitoring provisions whatsoever and, perhaps as a consequence, two-thirds do not delineate specific allocations and four-fifths have no enforcement mechanism. Moreover, those treaties which do allocate specific quantities, allocate a fixed amount to all riparian states but one – that one state must then accept the balance of the river flow, regardless of fluctuations.

While wars do not seem to be fought over water, water-tensions do crop up regularly which, if one considers each treaty the manifestation of a dispute resolved, are exemplified by the very prevalence of such treaties. By examining the conditions in each set of co-riparian states immediately before the resolution of their disputes, one finds patterns which not only may indicate future water dispute, but may offer clues to the type and intensity of an impending conflict.

These indicators might include water quantity issues, water quality issues, multiple uses of the water, political divisions, the geopolitical setting, level of national development, the hydropolitical issue at stake, water institutions, and depth of any national water ethos.

Notes

1 The material from the above section was drawn from a paper presented at a NATO Advanced Research Workshop on Environmental Change, Adaptation and Human Security, Budapest, Hungary, October 9–12, 1997.
2 This is not quite true. The earliest documented interstate conflict known is a dispute between the Sumerian city-states of Lagash and Umma over the right to exploit boundary channels along the Tigris in 2500 BCE (Cooper, 1983). In other words, the last and only 'water war' was 4500 years ago.
3 The India–Nepal Kosi River Project Agreements, signed in 1954 and 1966 provide two examples.
4 The 1951 Finland/Norway treaty and the 1952 Egypt/Uganda treaty both include such compensation.
5 In this regard see, for example, British colonial treaties and the 1947 Allied peace treaty with Italy.
6 'Power' in regional hydropolitics can include riparian position, with an upstream riparian having more relative strength *vis à vis* the water resources than its downstream riparian, in addition to the more-conventional measures of military, political, and economic strength (see Lowi 1993/95). Nevertheless, when a project is implemented which impacts one's neighbors, it is generally undertaken by the regional power, as defined by traditional terms, *regardless* of its riparian position.

9
Water and Conflict in the Middle East and South Asia
Miriam R. Lowi

In the academic and policy communities, the concern with security has traditionally focused on the threats that states pose to each other in a system of anarchy.[1] Security is about the ability of states and societies to maintain their independent identities, as well as their physical and functional integrity. The referent object of security is the state, hence the term 'national' security, while the concern with security engages inter-state relations and the threat of organized violence through military means and intentional actions. Over the past quarter century, and especially since the end of the Cold War, the significant decline in the threat of war among great powers and the disappearance of bipolarity have sparked an interest in new or previously overlooked threats. Well prior to the fall of the Soviet Union, scholars of international affairs were suggesting that our understanding of security needed to be revised and refined. Richard Ullman, in a seminal article, offered a useful definition: 'A threat to national security is an action or sequence of events that (1) threatens drastically and over a relatively brief period of time to degrade the quality of life for the inhabitants of a state; or (2) threatens significantly to narrow the range of policy choices available to a state or to private, non-governmental entities (persons, groups, corporations) within the state' (1983, 133). Because core values, the survival, and/or welfare of the nation are at stake, a threat to national security requires a centrally coordinated national response to mitigate or reverse. In this conception of security, both the types and the sources of threats are wide-ranging. Moreover, non-traditional threats engage non-military players in actions that are neither clearly intentional nor directed by the state. (In addition to highlighting the need to redefine security, some scholars and analysts have been advocating an extension of the referent objects of security to include, in addition to the state, the 'community', the 'region' and the 'planet', among others.)

In light of this conceptual shift, there has been a multidisciplinary debate in recent years, over whether or not changes to the natural environment should be considered national security concerns. The debate engages two fairly broad camps within the environmentalist community: The 'naysayers' argue either that the conventional language and militarized framework of security make the inclusion of environmental issues inappropriate strategically (Deudney, 1990), or that there is no empirical support for their inclusion (Levy, 1995). The 'yea-sayers' (Lonergan, 1993; Myers, 1993) maintain that all environmental issues are potentially security issues because security, despite the traditional state-centric focus, is in essence about the quality of life of individuals, which is bound to be affected by the depletion or degradation of environmental resources.[2]

Unfortunately, the term 'environment', just like the term 'security', is itself contested. A persuasive definition is one that is limited to 'biological or physical systems characterized either by significant ecological feedbacks or by their importance to the sustenance of human life' (Levy, 1995, 39). Thus, non-renewable natural resources, such as petroleum and other minerals, are excluded from this definition. However, their exclusion has less to do with their finite nature and more to do with two other criteria – having to do with function and manipulation – that distinguish environmental issues from resource issues. The first criterion has to do with whether the resource is primarily of economic value, as is coal, or is life-sustaining, as is water.[3] The second has to do with whether the resource is utilized through mining and extraction, or is subject to damage through human activity.[4] For both criteria, the former characterizes a resource issue, the latter, an environmental issue. This distinction, when it can be made, is an important one because environmental problems, such as pollution of a drainage basin, are not resolved in the same way as economic problems, such as determining the potential beneficiaries of a coal mine. Environmental issues, therefore, commonly refer to damage to the air, water or soil – hence implicating cropland, forests and marine life – in a manner that goes beyond natural processes and that usually results from anthropogenic alterations.

As one of the authors in a collaborative research project closely linked to this debate,[5] I feel called upon to react to some of the major issues in terms of my empirical research and its focus on one environmental resource: disputes over access to and control over scarce water resources in protracted conflict settings (Lowi, 1993, 1995). At first glance, the empirical material may appear to engage only distributional issues, and not 'environmental' issues as defined above. It is important to note, though, that insofar as water is concerned, the two are closely related: resource allocation issues, such as 'water rights', have important environmental effects. In the case of surface water, for example, the damming of a river upstream not only reduces the quantity of flow to downstream users, but can degrade the quality of the soil downstream as well, if the river flow falls below a certain minimum required to flush out salinity. Moreover, if water impounded behind a dam is used to irrigate soils and then that same water is released into the river system, downstream users receive water of inferior quality. Also in the case of groundwater, distributional and 'environmental' issues coalesce: within a single subterranean basin, for example, over-pumping in one area not only draws down the water table in other areas of the basin, but also, may provoke salinization and/or the intrusion of pollutants. Rarely are disputes over access to and control over water resources simple scarcity, or allocational, issues; more often than not, they have severe implications for the quality of the resource itself and, at times, related resources, such as land, as well.

This chapter is organized as a series of responses to a set of questions that have been the focus of the so-called 'environmental security' debate: first, should environmental change be considered a national security concern? Second, is there a positive relationship between environmental change and acute conflict?[6] Finally, are the effects of environmental change in different parts of the world so critical that, in the absence of immediate action, irreversible humanitarian disaster is inevitable?[7] My response to this last question is tentative, as its purpose is essentially to suggest new avenues for research in the field of environmental scholarship, and move away from the dominant, yet circumscribed, environment/conflict nexus. My responses to the two former questions are highly qualified and case-specific. Although this may cause readers some frustration, the suggestion is that we cannot generalize in such matters and that need not be problematic.[8]

In the above questions, the term 'environmental change' is used as a shorthand – and admittedly an imperfect one – for any of the following phenomena: resource depletion, resource degradation, and resource 'capture', or an altered and inequitable distribution of the resource.[9] A specific 'water-related' question is appended to each general question in order to give substance to the issues. It is important to note that although the responses are specific to water, they may be applicable to other environmental resources, as well. However, it would be imprudent to make generalizations in the absence of an array of case studies pertaining to changes to cropland, forests and marine life in a variety of settings.

The empirical evidence that informs my responses derives from work that I and others have conducted on the Euphrates, Indus, Jordan and Nile River basins, as well as on both West Bank and Gaza groundwater basins. All of the above cases concern shared, transboundary resources, and all but the latter engage inter-state relations directly.[10] In each case, water is scarce relative to demand in some part of, if not the entire, basin and degrees of supply-induced and structural scarcity persist, as well.[11] In all cases, environmental systems, whether surface or subterranean flow, have been degraded and/or are likely to be degraded by human activity such that inter-state relations are and will continue to be affected. Finally, all except for the Nile case are located in protracted conflict settings; the Nile case, however, has considerable conflict potential.

Environmental change

Should environmental change be included among the potential threats to a state's security? More specifically, is the depletion, degradation, or inequitable distribution of water resources a national security concern? The short answer to this general question is that it depends; it depends on the context. It also depends on the implications of the change, the extent to which the state was dependent upon the *status quo ante*, and the extent to which it is prepared to respond to the change relative to other (potential) national security concerns. Moreover, it is a matter of perception and of degree. There can be no *a priori* answer; the question can only be answered on a case-by-case basis.

In arid and semi-arid regions, for example, water scarcity relative to demand is not only a fact of life, but also, and more importantly, a material constraint to survival. Without unimpeded access to water resources, populations simply cannot survive, and states, bound as they are to protect their populations and guarantee their welfare, cannot pursue the multiplicity of tasks they are expected to fulfill: development of the economy, settlement of the population, provision of potable water, to name but a few. The survival of the state as both a physical and a political entity is dependent upon the provision of these goods.

Hence, if we define threats to (national) security as Ullman does, then it would be fair to say that under certain conditions, the depletion or degradation of the water supply could constitute a national security concern. However, the degree to which it is a concern is a matter of perception, of judgement, and of particular circumstance. Where both the water endowment and the number of consumers remain more-or-less fixed, there is considerably less cause for concern. If, however, we consider the case of an arid or semi-arid country with a high population growth rate, the threat of resource depletion or degradation is a constant source of concern precisely because of competition among an increasing number of (potential) users. The same is true when the water endowment is a transboundary resource, and/or the quantity, quality or timing of its flow is, for one reason or another, uncertain.

In international river basins, the degree to which the riparian states perceive water to be a national security concern depends not only on the degree of dependence on the shared resource relative to other exploitable sources, but also on geographic position within the basin – that is, the upstream/downstream configuration – and political relations within the basin. Geographic position is important because the upstream riparian is in the most favorable position. Being at the source of the river, it can, at least in theory, exploit the water as it wishes: it can diminish the quantity and the quality of water available to states downstream. The latter must suffer the consequences unless, of course, an international agreement governs water rights in the basin or the downstream state is militarily the more powerful. Political relations, no doubt, do make a difference. Where characterized by hostility, then, in the absence of an international agreement stipulating water usage, the resource in question is far more likely to be considered a national security concern than would otherwise be the case; this is especially so for downstream riparians who may perceive themselves to be at the mercy of those upstream. The conflictual environments of the Nile, Euphrates, Indus and Jordan basins provide substantial evidence for the importance of this combination of factors.

Consider the case of Egypt, one of 10 states in the Nile River basin. Except for the East African plateau and the Ethiopian highlands, the Nile basin is arid throughout. The portion of Egypt from its border with the Sudan to Cairo is considered an extremely arid zone. Moreover, for Egypt (as well as for the north of Sudan), the Nile is the only source of water for meeting consumption demand: the country's dependence on the river system is absolute. That, coupled with the marked seasonal variation in river flow, causes Egypt to suffer from insufficient water, especially during the very long, parched summer months. The creation of storage capacity behind the Aswan High Dam (1960–70) was essential for coping with the unpredictable flow.

As in the Jordan and Indus basins, and the Euphrates basin in Iraq, settlement in the Sudanese and Egyptian portions of the Nile basin is intimately associated with the river. This is especially striking in Egypt, where a population of roughly 63 million in 1996, with a population growth rate of 2.3 per cent per annum, is crammed into a habitable area of about 30 000 square kilometers, a narrow strip of land along the banks of the Nile and in the delta north of Cairo.

Furthermore, Egypt is the furthest downstream in the basin. The river system is composed of two major tributaries, the White Nile and the Blue Nile, which rise in Lake Victoria (Kenya, Rwanda, Tanzania, Uganda) and Lake Tana (Ethiopia) respectively. The two branches meet at Khartoum to form the main Nile, which continues northward through the Sudan and Egypt to the Mediterranean Sea. Fortunately for Egypt (and the Sudan), the six relatively well-watered, upstream states on the White Nile, including Burundi and the Democratic Republic of Congo (formerly Zaire), have shown little interest to date in exploiting the river. However, Egypt's downstream position relative to Ethiopia, at the source of the Blue Nile, is a cause for concern. The Blue Nile carries an average annual flow of 50 billion cubic meters of water to its confluence with the White Nile, representing as much as 60 per cent of the total discharge of the Nile system. Until recently, Ethiopia showed little interest in the hydraulic potential of the river. But because of population growth and declining food security, it needs to harness the abundant waters of the Blue Nile. Recently, in fact, both the Sudan and Egypt protested parliamentary debates in Ethiopia concerning unilateral action on that source.[12] Needless to say, Egypt has grounds for concern.

The case of Egypt highlights the variety of contextual features that help clarify whether and to what degree the depletion and/or degradation of water supply may be considered a national security concern.[13] These features include: (a) the quantity and quality of the resource endowment relative to (present and future) consumption demand; (b) the 'nature' of resource dependence; that is, whether the state is dependent on one or more sources and whether the source (or sources) is (are) shared, as in a transboundary river; (c) in the case of transboundary rivers, the number of riparians involved, the nature of relations with the other riparians, and finally, geographic position within the basin. For Egypt, aridity, high population growth, absolute dependence on one, shared body of water, downstream position, and the threat of important extractions upstream coalesce in a perception that water is indeed a security concern: deleterious changes to the water supply would threaten the country's welfare and ultimately, its survival, and would invite some centrally coordinated national response. There is, however, an important mitigating condition: Egypt's relative power in the basin – in terms of both military force and economic and political power – is superior to the other riparian states, such that the latter are unlikely to engage in actions that could provoke a hostile response downstream.[14]

Of the three riparians in the Euphrates River basin – Turkey, Syria and Iraq – Turkey is in the most advantageous position: it has several relatively abundant rivers and enjoys the greatest water endowment relative to demand; it is the upstream state in the basin and, especially since the Gulf War, it is militarily the most powerful. Both downstream states have extensive desert and semi-desert zones, comprising about one-half the land area of Syria and two-thirds of Iraq. The Euphrates accounts for as much as 86 per cent of the water available to Syria, the mid-stream riparian. As for Iraq, the furthest downstream, agriculture in all but the northern portion of the country is heavily dependent on water from both the Tigris and the Euphrates rivers.

Since the mid-1960s and in part to combat the irregular spatial and temporal distribution of water in the country, Turkey has been building dams on its portion of the Euphrates river. The *GAP*, or *Southeast Anatolia Development Project* is a massive water management scheme that involves dam building, diversions, and the extension of irrigated agriculture in the southeastern portion of the country.[15] Today, work on the project is well under way, and the planned withdrawal of 14–17 bcm (out of a total of 32 bcm) promises much hardship downstream. Syria's ability to generate hydropower will be curtailed by the depleted water levels. Its ability to extend irrigated agriculture will be hampered by both the depleted water levels and the inferior quality of water that is reintroduced into the system. Iraq, which must contend with the water engineering schemes of both upstream states, will have to forfeit part of its intake from the Euphrates and settle for water of inferior quality, as well.[16] The country will be hard-pressed to meet consumption demand in its part of the basin, and it will no longer have at its disposal the large volumes of fresh water with which to reclaim its already highly saline soils. Indeed, the key environmental effect of the *GAP* project is not so much the reduced flow – a distributional issue – as most of the water will be reintroduced into the system, but rather, the inferior quality of that flow. Because much of the water released downstream will have already been used for irrigation, it will have a relatively high content of salts and contaminants that will render it unusable for the growth of certain crops and will degrade the quality of soils along the river banks and surrounding areas.

Contiguity gives rise to shared interests and concerns, be they quiet borders, the reining in of a transnational secessionist movement, or undisturbed access to transboundary resources.[17] In the Euphrates River basin, the combination of adversarial relations, physical proximity, and resource interdependence magnifies the security concerns of the downstream states. The threat of resource depletion and degradation functions as a subset of their concerns. Furthermore, the hostility between Syria and Iraq precludes the possibility of bilateral cooperation in an effort to counterbalance Turkey.[18]

A similar combination of factors inheres in the Indus and Jordan River basins. In both cases, the downstream state – Pakistan in the former, the Kingdom of Jordan in the latter – is tremendously dependent on uninterrupted access to the waters of the transboundary river. In both, protracted political conflict governs relations among riparians and acts as a profound constraint on what can be achieved among them. When the Indian sub-continent was partitioned in 1947 to create two distinct political entities, the new international boundary cut across the Indus River, its 5 tributaries and canal systems that had been developed under the conception of a single administration. As a result, the headwaters of none of the rivers of the Indus system rose in Pakistani territory, and Pakistan became the lower riparian relative to India. Moreover, India had control over at least two tributaries that provided it with no irrigation water at all, but upon which (West) Pakistan was heavily dependent. (Lowi, 1993, 1995, 61–7). David Lilienthal, a former chairman of the TVA, described thus Pakistan's absolute dependence on the Indus waters, more than two-thirds of which originated in Indian-controlled territory:

> No army, with bombs and shellfire, could devastate a land as thoroughly as Pakistan could be devastated by the simple expedient of India's permanently shutting off the sources of water that keep the fields and people of Pakistan alive. (1951, 58)

Indeed, less than four months after partition, India suddenly cut off the flow of water to its downstream neighbor from two of the Indus River canal systems. This hostile act depleted the supply of irrigation water to West Pakistan at a critical time for sowing, and deprived the city of Lahore of its prime source of municipal water. Whatever the objective may have been, India was implicitly asserting upstream riparian proprietary rights. The precarious survival of newly-created Pakistan was such that the World Bank anxiously accepted to negotiate the 'Indus question'.

The *Indus Waters Treaty* was signed in 1960. Eighty-one per cent of the water was allocated to Pakistan and the remaining 19 per cent to India; Pakistan was given priority over the three western tributaries, India over the three eastern tributaries. Included in the treaty were stipulations for the building of storage facilities and link canals that would guarantee to Pakistan its historic withdrawals from rivers now under Indian control. In essence, the treaty extended the process of partition; through extensive diversions and redirection of flow, the rivers themselves were territorially divided between the two adversaries. In the aftermath of partition of the sub-continent, the concern for absolute sovereignty was paramount; hence, the final plan for the utilization of the Indus waters was formulated on the basis of no interdependence.

This solution highlights the significance of states' hierarchy of security concerns. While diversionary activity was bound to have important environmental effects in the basin, this was perceived to be secondary to the need to create two independent river systems. Indeed, environmental degradation – especially increased salinity and bacterial life in the water and surrounding soils – has affected the productivity of agriculture and the quality of drinking water.[19]

For Pakistan, however, its inferior power resources, downstream riparian position, and acute dependence on the Indus waters made it highly vulnerable to its adversary upstream. Moreover, the bitter history of relations between the two communities caused both India and Pakistan to view the very proximity of the other, as it was constituted, as a 'structural political threat' (Buzan, 1991, 121–2). In this context, concerns regarding the eventual degradation of the Indus waters, attendant to the hydro-engineering alterations of the river system, were less prominent, in the aftermath of partition, than were the immediate concerns for physical separation, territorial and political sovereignty.

The final case to consider is that of groundwater sources shared by Israel and the West Bank. The critical point to note is that approximately 40 per cent of Israel's sustainable annual supply of groundwater, or one-quarter of its total renewable fresh water supply, originate in Palestinian territory which, from 1967 until recently, was wholly occupied by Israel. These waters flow naturally across the 'Green Line' – the Armistice Demarcation Line established in 1949 – into Israel, where they are tapped and integrated into Israel's National Water Carrier system. Because of the importance of this source to Israel's overall water balance, from the beginning of the occupation of that territory in 1967 until 1995 and the agreements within the framework of the peace process, stringent measures were adopted and enforced by Israel's Water Commission Administration to ensure that the Palestinian population had access to no more than 20 per cent of the groundwater, thus guaranteeing at least 80 per cent to Israelis across the 'Green Line' and in settlements on the West Bank (Lowi, 1995, 125–7).[20] The objective of these measures was not only to fix allocations, but also to monitor their utilization and thus protect the quality of the supply earmarked for Israel.

The extent of Israeli dependence on West Bank groundwater highlights the perceived significance of the West Bank for the security and development concerns of the State of Israel, and exposes one of the major reasons why Israel is not easily relinquishing control of the territory. Without access to the rich groundwater reserves, Israel could be denied some portion of the water supply it has been consuming to date. Moreover, intensive pumping of aquifers on the West Bank could deteriorate the quality of groundwater that flows beyond the 'Green Line' into Israel. For the Palestinians, however, control over this 'national' resource is equally important;

the socioeconomic development of a future Palestinian state would require harnessing domestic water supplies. No doubt, it is because water resources and national security concerns are intimately linked in the Israeli–Palestinian arena that in the peace talks that were initiated in 1991, discussion of 'water rights' was put off until final status negotiations.[21] Yet the fact that the adversaries embarked upon a peace process may imply that they consider the ongoing hostility a far greater threat to 'national' security than are actual or potential changes to water supplies. To wit, it may be easier to mitigate threats to environmental resources once progress has been made toward resolving political conflicts.

To summarize, the depletion, degradation, or inequitable distribution of water resources tends to be a national security concern especially when the water in question is in short supply while dependence on it is great, and it is shared with other, often adversarial, states. Geographic position relative to the resource, political relations in the basin and the relative importance of the water issue to those relations, and the configuration of power within the basin significantly affect perceptions as to security implications. In the cases outlined above, and because of the very particular combinations of factors that inhere, environmental change has been, and continues to be, included among the potential threats to states' security.

Environmental change and conflict

Is there a positive relationship between environmental change and acute conflict? Specifically, has the depletion or degradation of water resources provoked war or violent confrontations between states? The short answer to this question is mixed; yes, there is a positive relationship between environmental change and acute conflict, but no, the depletion and degradation of water resources *per se* have not provoked inter-state violence. In the Euphrates, Indus, and Jordan basins, the conflicts have never been driven by environmental concerns. Especially insofar as the Indus and Jordan River basins are concerned, the conflicts are both multi-faceted and multi-dimensional. Environmental changes, specifically the depletion, degradation, or altered allocation of water resources, or the threat thereof, have functioned as aggravating features of the political conflicts, but in no case to date have the riparian states gone to war over water.[22] The environmental concerns of states tend to be secondary to their 'high politics' concerns.

Ismail Serageldin (1995), the Vice President of the World Bank, has said that the 'wars of the next century will be over water'. Politicians in the Middle East and scholars of the Middle East have made statements suggesting that all Arab–Israeli wars have been wars over water, and/or that future wars in the region will be over water. Both the 1967 War and the 1982 invasion of Lebanon, the episodes most often touted as 'water wars', demonstrate the inaccuracy of such statements. To begin, however, we must illuminate some features of the hydrogeography of the Jordan River system.

The Jordan River is formed by the convergence of three tributaries – the Hasbani, Dan, and Banias – that rise in Lebanon, Israel, and Syria respectively. The three meet in Israeli territory and flow southward into the Sea of Galilee (Lake Tiberias) and then into the Lower Jordan River. The Yarmouk tributary, which rises in Syria and forms the boundary between Syria and Jordan and then between Jordan and Israel, enters the Lower Jordan a few kilometers south of Lake Tiberias. The Lower Jordan, which flows into the Dead Sea, forms the boundary between Israel and Jordan for some of its distance.

In the late 1950s, the Israeli government began work on its National Water Carrier plan – a system of pipelines to pump fresh water from Lake Tiberias and transport it southward, outside the Jordan basin and throughout Israel. Once the plan became public and the Israeli government announced that the first stage of the diversion would be completed by 1964, the Arab states, and especially the Arab riparians, reacted vociferously. They perceived it as a violation of the rights of the Arab riparians and of those living within the basin, and a profound threat to the security of the Arab states (Lowi, 1993, 1995, 115–44). The Kingdom of Jordan, most dependent of the Arab riparians on the waters of the Jordan River and most likely to suffer from a depletion of fresh water in Lake Tiberias, made a strong plea for concerted action to prevent the execution of the Israeli project. A proposal was submitted to the Arab League calling for the diversion of the Hasbani into Lebanon, the channeling of its surplus waters to the Banias in Syria, and the diversion of the surplus waters of the Banias to the Yarmouk for the benefit of Jordan. The Arab League adopted the plan in 1960, despite the dissenting voice of Syria, which advocated a military response.

In 1964, within weeks of the commencement of work on the Arab diversion project on the Banias tributary, there was a border clash between Israeli and Syrian forces: the first in a series of military responses to rival water projects. Threats and counter-threats among the basin states and regarding the utilization of the Jordan waters were recurrent during this period. Indeed, throughout the first half of 1965, the Arab diversion was the cardinal political issue of inter-state relations in the Middle East. Israel's Premier Levi Eshkol stated that, 'water is a question of life for Israel', hence, any attempt to prevent Israel from using a portion of the Jordan waters would be considered a violation of Israel's right to exist. He warned the Arabs that, 'Israel would act to ensure that the waters continue to flow' into its territory. In keeping with its policy to consolidate its military strength, Israel made important arms purchases from Europe and the United States. Border incidents increased in number and intensity, and water installations were often the targets of attack.

By mid-1965, work on the Arab diversion in Lebanese and in Syrian territory came to a halt because of inter-Arab disputes; the diversion scheme was laid to rest. Nonetheless, the arms race, the flexing of muscle, and repeated border clashes between Israel and its Arab neighbors continued throughout 1966 and the winter of 1967. In May 1967, the Syrian government announced that Israeli troops were concentrating along Syria's southwestern border.

For both Syria and Egypt, the successive provocative statements made by Israeli officials revealed the intention to attack Syria. On May 18, 1967, Egyptian President Gamal Abd al-Nasir demanded the immediate and total evacuation of the United Nation's Emergency Forces stationed in the Sinai Peninsula. Expecting this would deter Israel from launching a punitive attack on Syria, he had the Egyptian army take up positions in the Sinai and along the border with Israel. Within a few days, Abd al-Nasir, responding to Jordan's taunts that he had abandoned Syria, announced that he was closing the Straits of Tiran to Israeli ships and to all ships carrying strategic cargo for Israel. Assuming that Egypt was preparing to attack, Israel launched a pre-emptive strike. By June 5, fierce fighting broke out between Egyptian and Israeli forces. Within no time, all states of the central Middle East got involved in this, the third Arab–Israeli war.

The Jordan waters crisis of 1964 and the Arab–Israeli war of 1967 were two distinct crises in a protracted conflict. Nonetheless, the former can be considered as one of several conflict spirals that, in combination, culminated in war. The 1967 war was neither over water, nor about water; to argue that would be to deny the 'richness' and multifaceted nature of the context of conflict in the Arab–Israeli arena. However, in the years leading up to war, water-related issues – specifically, the depletion, degradation, or altered allocation of water resources, or the threat thereof – were definitely part of the larger picture; at times they were in the foreground, at other times in the background. They were part of the package of issues that, as a package, established the conditions that made war likely.

While environmental changes *per se* did not cause the outbreak of war, they certainly were an important effect.[23] The geopolitical outcome of the 1967 war changed significantly the hydrogeography of the region: by occupying the Golan Heights, Israel became the upstream riparian on the Banias tributary and secured control over a greater portion of the Yarmouk waters. Moreover, its occupation of the West Bank meant that the rich groundwater sources fell within its jurisdiction.

In June 1982, Israel invaded Lebanon. Among the various explanations for that invasion, some analysts suggested there was a 'hydraulic imperative' (Stauffer, 1982). According to this view, Israel went into Lebanon to seize control of the Litani, a wholly national river that is located, at its southernmost point, within a few miles of the Israeli border. The veracity of this claim is difficult to sustain. Despite numerous anecdotes full of innuendo and considerable hyperbole, there is no credible evidence that the 'lure of the Litani' drew Israel into a long and costly war, nor that any efforts were made by Israeli forces during or following the invasion to divert a portion of the Litani waters into Israeli territory so as to mitigate problems of scarcity and enhance socioeconomic development. The far more plausible explanation for the invasion is the commonly acknowledged one: that Israel aimed thereby to smash the PLO, its institutions and power base in Lebanon, and in so doing, to deal a hard blow to the growing Palestinian nationalist fervor on the West Bank.

It may well be that for some elements of the Israeli political elite, the possibility of securing more water enhanced the desirability of an invasion. This, however, is no more than conjecture. To wit, during that invasion, Israel set up a 'security zone' in the south of Lebanon that includes the headwaters of the Hasbani tributary. Hence, by the summer of 1982 and by virtue of that military operation, Israel had assumed control over all three tributaries of the main trunk of the Jordan system. It was now the upstream riparian in the basin. That this was an aim of the invasion cannot be proven; that it was an outcome is certain.[24]

In sum, the depletion, degradation or altered allocation of water resources, or threat thereof, is never a necessary nor a sufficient cause of (violent) conflict. In no case that I am familiar with have states gone to war precisely because of changes to environmental resources upon which they depend. This, however, does nothing to diminish the significance of environmental factors as one among a number of causal factors leading to conflict. To wit, most conflicts can be traced to a variety of causes; and in conflict settings, environmental factors tend to function as intervening variables.

No doubt, the major wars of the first half of the twentieth century were inspired, to a degree, by the perceived need on the part of the German and the Japanese governments to mitigate resource constraints through 'lateral pressure' – by securing access to resources outside their recognized borders. In both cases, however, the resources sought after – petroleum, iron ore, tin, rubber – could, unlike water or air, be converted relatively easily into state power; all were essential for the conduct of military operations, in addition to being important ingredients of industrial development more generally. These were not environmental resources, as defined above; they were neither life-sustaining nor anthropogenically degradable. Furthermore, the resource component of those wars concerned only distribution; because the depletion and/or degradation of a resource base played no role, there was no 'environmental' component to conflict. In contrast, we have not found, to date, cases in which inter-state war derives primarily and predominantly from the depletion, degradation or inequitable distribution of environmental resources such as fresh water.

As with energy, territory and trade, states negotiate more-or-less successfully over environmental resources, with the goal of ensuring their independence and prosperity. States tend to cope with resource scarcity and (the threat of) environmental change, and they do so in a variety of ways: among them, technological innovation, trading,[25] and the establishment of regimes. Indeed, some of the most hostile neighbors, India and Pakistan, for example, have been able, against all odds, to formulate and implement workable and sustainable solutions to at least some of their water-related problems. Other hostile neighbors, such as those in the Jordan and Euphrates basins, have avoided using water as a natural weapon of mass destruction.[26] Adversarial states may threaten to take action – as the Israeli government did in response to the Arab diversion scheme in the 1960s, or as both the Syrian and the Iraqi governments did in the mid-1970s because of reduced intakes from the Euphrates River – but conflict over the environment does not naturally escalate and spiral out of control. This is especially the case with water because of the unique character of the resource – that is, its pronounced life-sustaining quality. Like air, the denial or degradation of water for human consumption would, if carried out beyond some threshold, qualify as a weapon of mass destruction.

The environmental concerns of states tend to be secondary to their political concerns. This is true, as well, in developing countries characterized, as they often are, by an unequal distribution of resources, pronounced competition among communities for access to those resources, and the systematic degradation of environmental resources largely because of intense competition among a constantly growing number of (potential) users. That there often is an environmental component to (both domestic and) regional conflicts in Third World settings is not surprising given the severe demographic factors, skewed resource endowments, and relatively weak institutional arrangements for establishing rights to resources and patterns of utilization. Environmental change tends to be part of the context of conflict; as is so blatantly clear in the Jordan River basin, the depletion, degradation, or altered allocation of water resources, in conjunction with other factors, create the conditions that make conflict likely. Their prominence as conflict-provoking factors varies in time, place and intensity, but they are never more than one of several components or contextual features of conflict.[27] Inter-state violence tends rather to result from fundamental and seemingly insurmountable disagreement over core values – identity, sovereignty, the survival of communities – and the distribution of power. In these 'high politics' conflicts, environmental changes function as intervening variables; they are aggravating factors, not determining factors.

Having said this, it is important to caution that the future may not mirror the past and the present (Kennedy, 1998; Wolf, 1997). Rapid population growth in the states of the Euphrates, Jordan, and Nile basins, and the developing world more generally, means that the demand for water will increase.[28] Existing supplies will have to be used more efficiently and allocated more appropriately to meet demand. If not, the needs of some communities for consumption and development will go unmet. The combination of high population growth and a more-or-less fixed resource supply may prove to be unsustainable and exceedingly volatile. Hence, the concern that 25 years from now, when the population of the Middle East has doubled, there could, in the absence of basin-wide cooperation, be wars – especially betweeen Israel and Syria, and Iraq and Turkey – to secure control over the headwaters of major rivers, is by no means unfounded. However, we need to be extremely cautious about making predictions.

Environmental change and humanitarian disaster

Are the effects of environmental change so critical that, in the absence of immediate action, irreversible humanitarian disaster is inevitable? Throughout history, the rise and decline of states and of communities have been determined to a considerable degree by the ability of those entities to adapt to and mitigate environmental constraints. In an effort to ensure socioeconomic progress, communities are forced to find ways to control population growth, distribute, develop and conserve resources, and thwart unsustainable environmental damage. Those that fail utterly in this task do not survive as coherent entities; those that succeed, at least partially, continue to persist. Success or failure hinges upon capacity[29] – the capacity to adapt, to find and implement appropriate and effective solutions, and to promote positive change.

We need only consider some of the bleaker environmental contexts to highlight the significance of and prospects for the generation of ideas to resolve problems. In the Jordan River basin, for example, water resources and arable land are scarce relative to current and projected demand. Moreover, hostile communities compete over access to supplies that are not only limited in quantity, but also, diminishing in quality largely because of competition in the midst of scarcity. The persistence of conflict in the basin prevents the communities from coming to mutually satisfactory arrangements regarding the distribution, management and conservation of resources.

This being said, during the past fifteen years, a wealth of studies has emerged from the region, from both the Arab and the Israeli communities. The studies outline proposals about how best to share, conserve and manage the resources, and augment supply. With regard to water augmentation, for example, there has been a flurry of research activity in the area of water transfers, resulting in a host of imaginative schemes to transport water from relatively wet zones to the arid and semi-arid Jordan basin (Lonergan and Brooks, 1994, 180–6; Lowi, 1993, 137). There also have been cooperative Israeli–Palestinian research projects in academic settings. Among the more thoughtful and creative proposals is one that identifies and raises options for joint management of the shared groundwater aquifers, and 'analyzes the institutional aspects of joint management structures' in a manner that could later serve decision-makers and negotiators (Feitelson and Haddad, 1995). Moreover, there have been numerous conferences, meetings and edited volumes that address concerns with regard to sharing, managing, and developing scarce water supplies.

Water is a scarce resource in the Middle East, but human ingenuity is not. Whether by themselves or in conjunction with others, the peoples of the Middle Eastern region will find creative and effective ways to enhance their resource endowment and meet their development needs. Nonetheless, the implementation of the various schemes for adapting to and mitigating environmental constraints awaits the right political context. In other words, the 'scientific' capacity to generate ideas and find solutions to environmental problems exists; the circumstances under which that 'scientific' capacity can be brought to bear effectively on environmental problems are lacking. This, I would argue, is the case throughout much of the developing world.

Poor countries may indeed be more prone than rich countries to 'market failures, social friction, capital scarcity, and constraints on science', but that does not imply that they suffer, as a result, from a shortage of ideas.[30] The condition of relative poverty poses material challenges and constraints, but it does not seal the fate of the poor.[31] Nonetheless, it is the case that in different geographic settings and for a variety of reasons ranging from political conflict to weak institutions to predatory governing structures, capacity and ingenuity cannot be mobilized to advance the security and well-being of states and peoples. What needs to be done, therefore, is not to try to anticipate impending disasters, as some of the more sensationalist environmental literature seems to do, but rather, to work towards enhancing the creative capacities of peoples, and equally, if not more importantly, to assist in developing the conditions whereby those capacities can be mobilized to improve human welfare.

Because circumstance may impede the application of human capacity and problem-solving ideas, we cannot rely systematically on these 'resources' to attenuate environmental stresses and strains. However, some degree of capacity and ingenuity can often be brought to bear; and it is for this reason that states do tend to cope more-or-less successfully with resource scarcity and the threat or reality of environmental change. Furthermore, in cases where states or communities are on the verge of total breakdown and threatened with annihilation, the explanation for their precarious condition lies not with the natural environment nor changes to it. Environmental factors may indeed exacerbate the precariousness of existence – again, they function as intervening variables. However, total breakdown tends to result from some combination of political conflict, weak institutions, and predatory elite behavior, often within a context of scarcity and inequity. In those settings, creative capacity cannot be effectively mobilized.

Conclusion

The empirical evidence illustrates that to varying degrees and depending on the context, the depletion, degradation or altered allocation of water resources may be considered a (potential) threat to the security of states and, as in the case of Israel/Palestine, communities. However, to date, changes to water resources *per se* have not provoked inter-state violence, although they have often been an aggravating factor in conflict settings. As such, they are a component of conflict, and usually one of several components. It is important to study the environmental component of conflict, just as it is important to study all components, since it is only with an appreciation of complexity that avenues toward conflict resolution or reduction can be addressed effectively. Nonetheless, since environmental factors are neither necessary nor sufficient causes of conflict between states or communities, it would be totally misguided to focus attention on resolving the environmental component either in an effort to reduce or resolve the larger conflict, or in the absence of efforts to reduce or resolve it.[32] Finally, the destruction, or threat of destruction, of states or communities does not derive, solely or even primarily, from environmental factors. More often than not, some combination of political, structural, and institutional variables is the primary cause. Hence, it is those variables that require our immediate attention. Furthermore, in many settings, political and institutional reforms would facilitate the attenuation of environmental constraints.[33]

We need to continue the enormous, but invaluable task of studying conflicts and their component parts. We need to understand why and how the various components emerge, develop, and in combination, provoke acute conflict. Equally, if not more importantly, scholars of the environment need to examine the nature of institutions and leadership, and the relationship of the state or the community to the external environment, since the capacity to respond effectively to constraints and enhance security and welfare depend to an important degree on these attributes of human collectivities. The task at hand is to promote capacity and problem-solving strategies, and work toward creating the conditions whereby they can flourish so as to effectively address and resolve the challenges that we face.

Notes

1 The author wishes to thank Jack Goldstone, Tad Homer-Dixon, Aaron Friedberg, Richard Matthew, Brian Shaw and Aaron Wolf for their helpful comments on earlier drafts. The paper on which this chapter is based, was delivered at the Workshop on Environment and Security at Columbia University on November 12, 1996 and at the Research Program on International Security at Princeton University on November 22, 1996. Sincere thanks to the participants in both series. A version of this chapter appears in Lowi (1999b); see also Acknowledgements, page ix.
2 The work of Thomas Homer-Dixon lies somewhere between the naysayers and the yea-sayers, but tends more toward the latter. Richard Matthew (1997) employs the terms 'rejectionists', 'maximalists', and 'moderates' to distinguish among the positions.
3 Note that water can have an economic value, in addition to being life-sustaining, but coal has no life-sustaining value at all. Like water, air has life-sustaining value, but no economic value to speak of.
4 The second criterion distinguishes between acts of god, about which we can do nothing, and human activity, about which we can do something (Shaw, 1996).
5 *Environmental Change and Acute Conflict*, A Joint Project of the University of Toronto and the American Academy of Arts and Sciences, 1990–92. The project set out to investigate the linkages between the depletion and degradation of natural resources and social strife and conflict in different parts of the world (see Lowi, 1992, 1993).
6 At issue are the linkages from processes of environmental degradation to the deterioration in security positions. Connections that run in the opposite direction – from the use of force to environmental change – are not examined here.
7 Although few scholars today address this question outright and consider it a legitimate prospect – Robert Kaplan (1994), Paul Kennedy and Matthew Connelly (1994), and Norman Myers (1993) being obvious exceptions – one does get the sense, from the sensationalism that characterizes some of the writing in this field, that some analysts are anticipating a human-induced environmental apocalypse. For that reason, I address the question here. For earlier concerns with the sustainability of the environment and the prospects for human survival see, among

others, Erlich, 1968; Hardin, 1968; Simon, 1981.
8 Another implicit suggestion is that there may be something wrong with the questions that define the environmental security debate, or at least that we should not fixate on them and ignore other, perhaps more pertinent questions.
9 Resource capture occurs when 'powerful groups within a society recognize that a key resource is becoming more scarce ... and use their power to shift in their favor the regime governing resource access' (Homer-Dixon, 1996).
10 Because of the uncertain status of the territories occupied by Israel in 1967, I suggest that the dispute between Israelis and Palestinians over the West Bank and Gaza groundwater basins directly engages inter-communal, rather than inter-state or intra-state, relations.
11 Note Homer-Dixon's classification of scarcity as: (1) demand-induced, when the result of growing population and an increase in per capita resource consumption, (2) supply-induced, when the result of resource depletion or degradation or, (3) structural, when resources are unequally distributed within society. See, *The Project on Environment, Population and Security*. Peace and Conflict Studies, University of Toronto, 1996.

Given that the demand for water in many parts of the globe exceeds its supply, water is, for functional purposes, a peculiar type of renewable resource. Hence, insofar as the 'environmental security' debate is concerned, the prevalence of supply-induced and structural scarcity may be more relevant than demand-induced scarcity.
12 I am grateful to John Waterbury for calling my attention to this recent development.
13 The case of Egypt also brings to the fore some of the concerns of other regional actors and of the United States government. Because of its critical position in the region both politically and strategically, a destabilized Egypt could call into question the stability of neighboring states and/or the region.
14 The other mitigating conditions are that: (a) Egypt has a storage facility on the Nile; hence, it is able, to some (albeit limited) degree, to protect itself from the variability in rainfall and river flow; and (b) there is a *Nile Waters Treaty* (1959) that, although outdated, in need of revision, and limited to only two of the ten riparians – Egypt and the Sudan – sets something of a precedent for future water agreements.
15 The project calls for the construction of 80 dams, 66 hydroelectric power stations and 68 irrigation projects covering up to 2 million hectares of land, at a total cost of more than $20 bill. (Kolars and Mitchell, 1991).
16 Daniel Hillel (1994) contends that Iraq may eventually lose as much as 80 per cent of its Euphrates inflow.
17 There is no reliable evidence suggesting that the Syrian government's support for the PKK, the Kurdish Workers Party in Turkey, is part of an effort to frustrate completion of the GAP project. Nonetheless, there is evidence that in recent years, the Turkish government has threatened to reduce the flow of Euphrates water to Syria, if Hafez al-Asad continues to support the PKK (Lowi, 1995, 138).
18 In addition to the fact that the political context in the basin itself

does not encourage negotiations, there has been very little external interest in mediating the Euphrates waters dispute and/or promoting its resolution.
19. Note that environmental degradation in the basin is not caused solely by diversions related to the Indus Waters Treaty. For further discussion see, Peter Gizewski and Thomas Homer-Dixon, 'Environmental Scarcity and Violent Conflict: the Case of Pakistan', the *Project on Environment, Population and Security*, Peace and Conflict Studies, University of Toronto, 1996.
20. Note that while some of the measures have been relaxed, the distribution of water resources has remained roughly the same in the latter half of the 1990s as it was prior to the agreements reached in the fall of 1995 (Lowi, 1996).
21. Again, this is not simply an allocational issue, as the term 'water rights' would suggest, since the quality of the supply is related to its distribution.
22. This begs the question of whether states are likely to go to war over water in the future. Population growth trends would be an important factor to consider in addressing this matter. See pages 164–5.
23. I am not suggesting here that environmental changes cannot be both a cause and an effect of conflict. No doubt, causal arrows may proceed simultaneously from processes of environmental degradation to deterioration in security positions, and from the use of force to environmental change. I am simply highlighting the significance of the latter to the Middle Eastern region in the aftermath of the 1967 war.
24. As in the case of the 1967 War, the more prominent causal arrows in the invasion of Lebanon run from the use of force to environmental change, and not from environmental change to the use of force, which is a central concern of the 'environmental security' debate. Similarly with India and Pakistan: the acute conflict between them provoked the eventual degradation of water resources and agricultural land. See page 157, above.
25. A logical solution to the condition of scarcity in the Euphrates River basin, for example, would be the trade of oil for water between Iraq and Turkey (Lowi, 1995,144).
26. Why this has been the case deserves considerable attention. I offer some preliminary explanations in Lowi, 1993, 1995, Chapters 3 and 9.
27. This does not mean that water issues are any less important as objects of study. The aim here is simply to gain a more accurate perspective on the role of water in conflict settings so as to be better equipped to address concerns with regard to conflict resolution, environmental rescue, and human security more generally.
28. For water demand forecasts in the Middle East for the year 2025, see Rogers, 1994, 307.
29. There is a large and growing body of literature that treats the issue of 'state capacity'; much of it concerns the capacity of regimes in the Third World to promote socioeconomic development. See, for example, Evans, 1995. For an effort to explore the relationship between state capacity and environmental change, see the *Project on Environmental Scarcities, State Capacity, and Civil Violence*, sponsored by the American

Academy of Arts and Sciences and the Peace and Conflict Studies Program at the University of Toronto.
30 Thomas Homer-Dixon argues that these four factors 'can limit the supply of social and technical ingenuity' (1995, 598).
31 See the provocative work of the eminent development economist, Albert Hirschman, who has argued that tremendous possibilities can be discovered in conditions of, even severe, constraint. Through rich case material, Hirschman illustrates ways in which communities in impoverished environments have been able to transform material limitations to their advantage; inter alia, 1971 and 1982.
32 This may be one of the hidden, and inadvertent, dangers of the research agenda that concerns itself with the environment/conflict nexus: that it may be construed as advocating such impracticable and inefficient conflict resolution strategies.
33 For interesting research along these lines, see Richards, 1983 and 1992, and Thompson (this volume).

10
Perceptions of Risk and Security: the Aral Sea Basin

Galina Sergen and Elizabeth L. Malone

Environmental security and environmental risk are closely related concepts. We argue in this chapter that researchers and policymakers can employ the approaches used in risk analysis to address concerns about environmental security. Environmental degradation most often engenders societal controversy, not armed conflict. Indeed, the potential for armed conflict – a focus of research in environmental security – is slight, and the literature emphasizing possible resource wars distracts attention from the need to study ways in which risk analysis can help to settle environmental controversies.

We first look at the meanings of security and risk, then define the three major approaches to risk assessment. Using a case study of water resources in Central Asia, we examine the existing controversies and the potential for armed conflict. We find that a focus on armed conflict is a narrow and unrealistic analysis; a more fruitful direction points toward the stability of sub-national institutions and equity concerns, that is, how fair any proposed action is to diverse groups in the region. With the latter focus, we can begin to understand how risk and security are socially mediated through meso-level institutions. We propose a risk-oriented framing that emphasizes pluralism as an essential element of environmental security.

Our intention is to broaden the debate beyond the academic realm of political science and to place the debate where it belongs, in disciplines that study and cope with environmental issues and their impact on society. Our primary assumption in this study has been that the very purpose of discussing environmental security is to understand all the dimensions of a potential problem in order to create stability and peaceful outcomes.

The meaning of security

Since the end of the cold war, the definition of 'security' has undergone close scrutiny and various attempts at redefinition (see, for example, Levy, 1995). While the literature on what security entails in the post-cold-war era is increasing, consensus is hard to come by. Perhaps the only point on which most authors agree is that this era has brought a new set of non-traditional security problems to the fore (Mathews, 1989; Gleick, 1991; Dalby, 1992; Sandholtz et al., 1992; Bedeski, 1995; Del Rosso, 1995). However, a decade after the dissolution of the Soviet Union, despite calls for a new, better, more acceptable security paradigm, a survey of the literature on the subject shows many competing paradigms, disciplines on different tangents, and no clear consensus as to what the term 'security' means. At the heart of the debate is that different people use their accustomed frames of reference to define security and confront security issues in different ways.

Some authors have adhered to a framing of security that limits the issue to military concerns – or those situations that could become the basis for violent conflict. Other authors have called for security (national, international and regional)[1] to be interpreted more comprehensively (Mathews, 1989; Sorenson, 1990; Lang, 1995; Dalby, 1992) and for the economic, environmental and ethnic aspects of security to be examined more fully. However, as the concept of security is broadened, the ambiguities are likely to increase. The real question is not how broad or narrow a definition should be, but what the implications are of any definition. The *American Heritage Dictionary* defines security as:

> 1. Freedom from risk or danger; safety. 2. Freedom from doubt, anxiety, fear; confidence. 3. Something that gives or assures safety, as: a. A group or department of private guards. b. Measures adopted by a government to prevent espionage, sabotage, or attack. c. Measures adopted, as by a business or homeowner, to prevent a crime such as burglary or assault. . . .

The positive and negative connotations imply very different emphases in environmental security. A focus on negative connotations – preventing espionage, sabotage, or attack – yields an emphasis on the absence of violence or crime. A focus on positive connotations – confidence and safety – yields an emphasis on a strong, resilient society. Environmental security investigations can focus on the potential for violent conflict or on factors that create societal stability. The emphasis on violence and crime is consonant with a framing of the issue as one of traditional public policy institutions, such as the military and intelligence organizations of nation states.

None of the case studies discussed by researchers in this area have proven that environmental degradation has been the sole or primary cause of violent conflict. By the time one arrives at the end of the logical chain – violent conflict – so many intervening variables have been added that it is difficult to see the independent contribution of environmental degradation (Levy, 1995, 58). The incompatibility of environmental concerns with violent conflict has been taken up by Daniel Deudney (1991). He argues that the case for linking environmental concerns with conventional military understandings of security is flawed, on the basis that military solutions are focused, technology-based, social interventions and that environmental problems are very often not given to such straightforward definitions.

There is also something rather sensational about linking the environment with violent conflict. Levy (1995) criticized scholars immersed in the environmental security discourse for using the word 'security' purely as a rhetorical device. Homer-Dixon (1996) agreed with him and, without considering his own work to be subject to the same criticism, accuses other scholars of using the word for the purposes of headline grabbing. He justifies his team's focus on the sensational links between environmental stress and violence on the basis that it 'helps bound our research effort' (1996, 49). Homer-Dixon's research is bounded to the point that it unfortunately misses the most interesting links between environment and security concerns. When a security issue that involves the environment is seen as the potential for armed conflict, only one, narrow line of research is possible, focusing on military action and excluding environmental and other non-military threats. When a security issue is perceived as a threat to stable relationships, and security concerns are the avoidance of violent conflict, research subsequently becomes more interesting and rich in possibilities. The research directive is to uncover the stabilizing influences that mitigate threats to security. Framed in this way, environmental security harmonizes well with the concepts of risk analysis that emphasize that people can avoid or mitigate undesirable events through choices they make (Renn, 1992). For both environmental security and risk, 'no single answer can be given to the problem of adequate safety in a complex society which contains a wide variety of perceptual biases about danger, expectations of the good life, and levels of demand for safety' (Rayner, 1987, 207).

The meaning of risk

Our view of the world and the risks it contains today is radically different from the societies of ancient Babylonia and Mesopotamia whose main concerns lay with crop survival and loan maintenance (Covello and Mumpower, 1985). While those concerns are held by modern farmers, risks in the farming industry alone have expanded and can now encompass pesticide use, introduction of new biotechnologies, rapidly changing markets and rising costs of production. However, the social and governmental responses to risk that are utilized today in risk situations such as recourse to compensation, insurance, and law, arise out of a collective wisdom derived from trial and error over centuries (Covello and Mumpower, 1985). In the legal systems of Britain, its Commonwealth and the United States, laws governing the procedures of liability and consent derive from the Code of Hammurabi and the Old Testament, and have been adapted over time in response to changing functions of the law (Hammer, 1980). Both the law and the economic mechanisms in place for compensation can be seen as practical responses to uncertainty, arising out of a need at the time of their development to reduce the burden of risk to the stakeholder so that progress and benefits such as trade, exploration and the acquisition of land and resources could proceed for the collective good.

The concerns that have arisen surrounding modern technological development have been coupled with an increasing concern for the preservation of the environment. Risks to humans and to nature have combined to increase the prevalence and importance of risk analysis. Risk analysts and policymakers need to account for the social aspects of risk decision making as they design procedures to scrutinize risk problems.

One dimension of environmental risk comprises the elements of uncertainty and indeterminacy in new technologies. People are increasingly wary of new technologies because of the risks of accidents such as those at Three Mile Island[2] and Seveso.[3] New 'wonder – drugs' – an example is thalidomide[4] – may carry their own risks, including deformation. More recent chemical and nuclear disasters, such as Bhopal,[5] Chernobyl,[6] and Sandoz,[7] have caused widespread and persistent problems, including public skepticism and distrust.

Another dimension of environmental risk is hazardous and toxic waste. By the year 2000, the US Government has agreed to reduce its current stockpile of chemical weapons as part of its obligations under the International Chemical Weapons Convention (Bradbury, 1994). In a series of studies and cost estimates, the US Army responsible for the weaponry has proposed that on-site incineration of the amassed stockpile is the most effective and efficient means for the US Government to meet its international obligations (Bradbury, 1994). Concerned about the safety implications of the incineration as well as the management style of the US Army, the local communities based near the eight proposed incineration sites in the United States have responded by taking their concerns directly to their Congressional and State representatives and the courts (Bradbury, 1994).

The case of the US Army, and the concerned local citizens who live near the proposed incineration sites, illustrates the way in which controversy can escalate, without open conflict arising. The US Army believes it has followed all the necessary regulations, while the citizens complain that their wider concerns about incineration have not been addressed (Bradbury, 1994). The arguments from both sides have been backed by sound scientific studies claiming varying degrees of risk to the local population. As a result, policy-making is informed by a number of contradictory studies and arguments, all apparently based upon good scientific analysis of risk. It often seems as if the more science is used to help support such social decisions, the more controversy is generated (Robinson, 1992).

The solutions for controversial decisions such as the incineration of chemical weapons are obviously not found in science alone. They rest with the perceptions held by all those involved and the degree to which opposing perceptions are respected and built into a possible solution. There are different tools available for assessing risk; there is not wholehearted consensus among the disciplines that study risk. Nonetheless the dialogue has been going on for over twenty years. This dialogue has examined some of the issues raised regarding environmental security. We discuss below the three general approaches to risk assessment relevant to environmental security issues; they are: technical risk assessment, psychometric or strategic risk assessment, and social risk assessment.

Probabilistic technical risk assessment

This presents risk as a standard formula:

$$\text{Risk} = \text{Probability} \times \text{Magnitude of Loss}$$

Developed during the 1940s, technical risk assessment involves three principal components: the threat of possible unwanted consequences and loss; the uncertainty that surrounds the nature of the threat (Rowe, 1977); and the opportunity of potential benefit. This is sometimes written as an equation:

$$\text{Risk} = \frac{\text{Probability} \times \text{Magnitude}}{\text{Time}}$$

Some risk outcomes may present benefits that are worth enduring some burdens; then again, the burdens may be latent and when expressed, may dramatically cancel out any benefits that were to be enjoyed. For example, climate change can be beneficial to growing certain crops in the mid-latitudes but cause out-migration from already marginal (hot and dry) areas, causing societal disruption that is a greater burden than the productivity gain. When it is isolated from social and political contexts and implications, risk can be viewed narrowly as a purely technical problem to be measured objectively and ordered in terms of probability and magnitude of benefit or burden; risk is not perceived as a social construct in the political arena of diverse world views. The underlying nature of this approach toward risk situations is strictly technical. The scientific 'experts' formulate the extent of the risk and demonstrate to the lay public how that risk should be managed (Nelkin and Pollack, 1980).

Psychometric/strategic risk assessment

Here, risk is regarded as a factor external to society. If citizens (such as those in the US Army chemical weapons controversy) object to the expert risk assessment or oppose the technical solutions, they are assumed to have psychological barriers that prevent them from recognizing what the real level of risk is. The strategic view of risk involves trying to correct public misconceptions about risk. By using systematic techniques of risk communication and establishing forums for discussion between expert and lay stakeholders, the strategic view attempts to correct the misperceptions held by the lay public about the true nature of the risk concerned (Rayner and Cantor, 1987). Strategic risk assessment sets out to grasp the fundamental differences between the expert and the lay stakeholder of a risk situation, with the specific mandate of finding ways to correct the perceived problem. The development of the strategic approach meant recognizing that there is a social perspective absent in seeking a purely technical solution, and that this will have implications for the social acceptance of risk policy implementation.

The social approach to risk

This presents a third view significantly different from the previous two. Risk, within this approach, is analysed as a social construct; that is people assign meaning and significance to possible events based on the social groups and institutions they belong to. Risk, in this view is not only a technical or physical problem. The model of risk communication favored is participatory as opposed to the closed or linear, top-down model often evident in the strategic and technical approach (Bradbury, 1993). Equity is one of the main concerns explicit in the social approach to risk assessment: equity in the distribution (and redistribution) of goods and equity in the social processes used to make decisions. The social approach assists stakeholders in a risk conflict situation to assess their requirements for fairness. The social approach to risk implies a participatory style of conflict management; that is, concerns for equity are just as important as technical analyses in arriving at satisfactory decisions. A wide plurality of views can be accounted for within the analytical frameworks that are used in this approach. Used in conjunction with one another, the different approaches outlined above have the ability to reach fair and well considered solutions to risk controversies. As the following case study shows, the appropriate style of risk assessment depends on the scale of the physical environmental problem, and the scale of controversy that surrounds it.

Controversies not conflict

To be considered a point of controversy, an environmental issue must be recognized as a tangible, physical problem affecting a given population. When the affected people choose a course of action to mitigate the problem, they will utilize their perceptions of safety, fairness, and the severity of the problem to guide their decision-making processes. In any populated area on the globe, and many unpopulated areas, there are the signs of anthropogenic change affecting the environment in various ways. Our expectations and understanding of what is good for us and for the environment vary greatly, perceptions which are culturally prescribed and conditioned by those around us and the infrastructure available.

In 1991, the republics of Central Asia – Kazakhstan, Kyrgyzstan, Tajikistan, Turkmenistan, and Uzbekistan – gained independence after the collapse of the Soviet Union. There had been no nationalist uprisings to force separation from Moscow, and independence was greeted in the republics with varying levels of uncertainty and hesitancy (Allison, 1996). With independence came increased attention from the West. Central Asia became known more widely as home to two man-made ecological disasters: the desiccation of the Aral Sea, and nuclear testing in Kazakhstan.

The Aral Sea in Central Asia is fed by two major rivers, the Syr Darya which flows from the Tien Shan mountains in Uzbekistan and Kyrgyzstan, and the Amu Darya which flows from the Hindu Kush mountains in Tajikistan. Kazakhstan and Uzbekistan border the Aral basin. The Syr Darya is shared by Kazakhstan (primary lower riparian), and Kyrgyzstan Tajikistan and Uzbekistan (upper riparians), while the Amu Darya is shared by Afghanistan, Uzbekistan, Turkmenistan (lower and middle riparians) and Tajikistan (primary upper riparian). The relationships between the Syr Darya and Amu Darya riparians are complex because the rivers define the boundaries and in some areas crisscross backwards and forwards between one republic and another.

During the 1950s, with the advent of pesticide development and Moscow's desire to increase production in Central Asia, decisions were made that would lead to the shrinkage and contamination of the Aral Sea. To supply the expanding cotton industry in the region, Moscow's central planners increased the amount of water drawn from the Amu Darya and Syr Darya rivers in Central Asia for irrigating the crops. Together with the construction of the Kara Kum canal[8] in Turkmenistan, these agricultural pressures and big industry priorities led to the catastrophic desiccation of the Aral Sea basin. Once the fourth largest inland sea in the world, the Aral has lost over one third of its area since 1960. The sand storms that sweep over the exposed sea-bed deposit large quantities of toxic dusts on agricultural lands around the basin.

Although there can be little doubt about the role of unwise agricultural and industrial policies contributing to the decline of the

Aral Sea and Amu Darya, there is also historical evidence to suggest that water levels in Central Asia, from the Pamirs to the Caucasus mountains, have fluctuated considerably through the centuries. Herodotus (c. 484 BC) described the Amu Darya, known in ancient times as the Oxus, as flowing due west into the Caspian Sea, avoiding the Aral depression. Aristobulus, a geographer of Alexander the Great, is quoted by Strabo (c. 63 BC) as saying that in the fourth century BC, the Oxus was used by Indian traders who made their way by boat on the Oxus to the Caspian, then the Black Sea, and so to Europe. Prince Alexander Bekovich, on a reconnaissance mission in the region for Peter the Great, heard from a Turkoman chieftain that the original course of the Oxus had been altered by a series of dams constructed in order to send the river toward the Aral depression (Hopkirk, 1994). The local populations during the nineteenth century were considerably smaller than they are today, yet they managed, in both their nomadic and settled lifestyles, to adapt to the changes and flourish in the region; maintaining, as always, their traditional irrigation schemes.

In 1992, a dam was constructed between the smaller northern part of the Aral Sea in Kazakhstan and the larger part of the sea that is situated mostly in Uzbekistan. The increase in the flow of the Syr Darya at that time, therefore, only benefited the northern sea in Kazakhstan until the dam collapsed and a small stream was allowed to flow from north to south. The Russians and the Kazakhs have attempted to rebuild the dam and to seal off the northern Aral completely from the southern part of the sea (Aladin, 1995). The most productive response to the environmental situation is through changes in land use and expectations about yield, but this is understandably easier said than done. To date, there has been considerable resistance to such changes. Potential strategies for mitigation are to improve access to clean water, and improve irrigation systems which would reduce the amount of water wasted each year. Projects from the United States Agency for International Development (USAID), the World Bank and United Nations organizations such as UNEP have been proposed to achieve these goals before the situation around the Aral Sea gets worse.

Hyman (1996) has suggested that if the economic and social situation within the republics were to deteriorate, governments may welcome disputes and conflicts with their neighbors as a distraction from the problems at home. Shirin Akiner wrote the following about inter-riparian relations around the Aral in 1993:

Hostilities are likely to become more acute as political and economic rivalries between the republics increase. The poorer mountain republics are already threatening to use water as an offensive weapon (by opening the dams to flood the plains, by blocking the flow so as to starve the downstream users, or by contaminating the flow) in disputes with their richer more powerful neighbors on the plains.

Such threats should be taken seriously; were they to be carried out, they would affect millions of people, and it would take years for the region to recover. However, in a desert region where water is scarce, we can assume that water is viewed and valued differently from areas where the resource is abundant. Hence, we can also expect that negotiations, arrangements, and equity issues will be brokered differently. It is not a foregone conclusion that political and economic rivalries in Central Asia will cause acute hostility, and the environment as a cause of war in Central Asia has yet to be thoroughly examined. The current approach to examining potential environmental security issues is illustrated by this 'worst case' scenario for the Aral Sea region by Stefan Klötzli (1994):

> Acute conflicts over the distribution of water and soil resources are likely to occur on the subnational level within irrigation systems existing along conflicting ethnic or tribal fault lines, together with other stress factors like economic depression, high rural population density, and unemployment.

Unfortunately, these statements about likely conflict are still focused on predicting the possible outcome. The factors for judging why a situation is more likely to result in conflict than in negotiation are not clear in the current literature. Research in this field sorely needs to move beyond the descriptive and examine why these claims may be true, if indeed they are. Would these subnational conflicts arise because the strategies for coping with water shortage and poor water quality between different social groups vary? Or would they arise because the solution for one group would deprive another? Given the history of ethnic conflict in Central Asia, how is this reflected in the modern social structure, the institutions, and degree of political representation given to the different groups? What are the public perceptions of the environment that influence choosing one strategy (violence, cooperation, economic bargaining) over another in order to cope with water and other environmental problems? How would the potential for conflict be evaluated, as opposed to the likelihood of negotiated settlement?

These are questions that would need to be asked if a social risk assessment were applied in the Aral Sea region. Regional and local institutional arrangements – their histories, strategies, inequalities – need to be clearly understood before equitable solutions can be implemented and sustained. They are the interesting background questions that take time and fieldwork to uncover. They are questions that are rarely, if ever, asked in debates regarding environmental security. This is because in the current literature there has been a tendency only to describe what *may* happen, rather than to uncover *how* and *why* these controversies may escalate. The reluctance to ask these more difficult and complex questions is the root of the problem with the environmental security discourse. The literature points to a disturbing trend that arguments must be legitimated by showing first that the problem is critical, and therefore worthy of attention and funding. So many of the discussions about the environment causing war are left to paint many unfortunate war scenarios that could, in certain specially prescribed circumstances, take place around the globe. It is a case of putting the cart before the horse. The notion that only violent conflict is worthy of attention, or that it should be used by academia as a topic to grab attention, is an unfortunate and outdated mindset.

When we looked at some of the suggested solutions for the Aral Sea crisis, and then looked into some of the views that have been expressed about water quality and availability by the most severely affected social groups close to the Aral, we found a variety of different strategies and viewpoints. To date, response to the Aral Sea problem from the national level has resulted in proposals to divert waters from the Siberian rivers and from the Volga. These schemes, which were proposed by the Russian government in the mid-1980s, reveal that the same impulses behind 'think big' style projects that led to the desiccation of the Aral Sea in the first place, characterize the potential solutions. Although these projects appear to have gone out of favor, the mitigation strategies are moving at a painfully slow pace. The problems that have emerged for the local populations around the Aral have distracted them from fighting to save the sea. Getting enough food and water to survive is a daily chore; this, no doubt, changes perceptions and priorities. For those severely affected, there is neither time nor resources to organize change.

There is plenty of room for controversy over the situation in the Aral Sea basin, but to date the problems have remained just that – problems, not direct conflicts. There is controversy over what to do about the Aral Sea crisis and how to mitigate its effects on local populations. Public expectations, priorities and perceptions of the problem about water arrangements and quality play a part in keeping things much as they are, without escalating tensions. One example of such can be found in northern Turkmenistan, where the quality of water available for drinking and sanitation is generally poor, but public perception of the problem reflects the fact that there are other things to worry about. Quality of household water has ranked second in concern to simply getting basic foodstuffs. A World Bank report notes the following:

> Interest in water quality far exceeds interest in sanitation and hygiene, and concern with water hardness and salinity exceeds concern with the bacteriological quality of the water, which is extremely poor. (World Bank, 1997)

Disputes over the nature of problems with water quality and availability in Central Asia will no doubt be rectified over time. Central Asia's independence opened it to the world in a way that had not been possible under Soviet rule, and the Aral Sea problem received heightened publicity in and outside the region when western scientists were able to see first hand the scale of the damage. But now, imposing our Western standards and expectations, and trying to direct the course of action from our own 'educated' perspectives, may not be helpful. If a strong civil society is to flourish in the region, bringing with it increased stability (Rayner and Malone, this volume), it needs to come from within the groups of Central Asia. Having gained independence from Soviet rule, the region needs to avoid dependence on Western aid.

One of the first signs that society was changing in Central Asia was the rapid formation of non-governmental organizations (NGOs), begun by people who had concerns with everything in their society from children's education, to environmental issues, to supporting small business ventures. These are positive signs that the republics of Central Asia are grappling with a difficult legacy and the pangs of new independence in a direct and so far resilient, manner. There are ways that are being sought to foster the growth of civil society in Central Asia, and these are important steps for the societies to take as they become more effective in making decisions that influence their immediate environment and their futures.⁹

Overall, Central Asia is riddled with environmental problems, from radioactive waste left by Soviet testing during the Cold War, to soils contaminated and leeched of nutrients from years of farming with heavy pesticide use and inefficient irrigation (Allison et al., 1996). With all its problems, Central Asia should, according to the predictions of the environmental security community, be mired in ghastly conflicts like its neighbor to the south, Afghanistan. Yet, the local populations are proving remarkably resilient and resourceful. There is still cause for grave concern about the myriad of environmental problems faced by the region. However, if strategies for improving the local populations' resilience to the risks (water contamination, lack of access to water, pesticide contamination, etcetera) are implemented and economic pressures do not reach breaking point, Central Asia will hopefully be the proving ground that while environmental problems are the cause of much controversy and hardship for those affected, they do not have to lead to armed conflict.

The situation in Central Asia could become complicated if the solutions to the environmental problems favor one group over another and are perceived to be unbearably unfair. There is a risk that conflict could occur over water pricing agreements, where the burdens of cost and responsibility, if not carefully negotiated and considered, could fall on one republic more heavily than another. Technology choice, as well as burden sharing, are critical considerations in reaching a viable and sustainable political solution. Groups must be able to afford the technology they choose; the repair, operation and future of the technology must be under the control of the group not dependent on outside assistance.¹⁰ Understanding community concepts of what is fair and what is not, can guide the policy maker's negotiation style and assist in reaching an equitable outcome.

What's wrong with this picture?

Two framing issues emerge from looking at the Central Asian situation. The first is that identifying, for example, a riparian issue in terms of its potential for conflict misses the whole range of ways that people solve their problems. The second is that a focus on the nation state not only leads to a discourse that dwells on and predicts armed conflict, but sets up a unit of analysis that does not represent the complexity needed to address an environmental issue. A focus at the meso institutional level and the use of risk analysis will lead to a more productive definition of the issue and candidate options for resolution.

Environmental security, then, cannot be limited to situations framed as nation-states avoiding war. First, environmental problems arise because human institutional arrangements have resulted in pollution or degradation (the cotton industry in Central Asia); therefore, various institutions (the collective cotton farms, *Kolkhoz*, around the Aral basin, designers and builders of the irrigation systems, etc) not just nation-states, are involved in security issues. Second, the stability of local and regional institutions is the critical factor in resolving environmental issues and maintaining regional stability.

Risk and security

The environmental security debate is suffering from being handled far too narrowly. To really understand how environmental controversies affect communities, which is the first step to understanding how they might escalate into something worse, the environmental security literature should contain fewer descriptions of 'what might happen if. . . .' and use interdisciplinary tools such as those used in risk analysis, where the goal is to avert controversy where risks threaten communities and their environment.

Environmental risk issues can be approached in a number of ways. The first phase is usually an assessment of the given physical system that is under threat, such as the Aral Sea and its desiccation. A technical risk assessment that includes the effects of the environmental threat on human health and life quality issues is devised for the geographic region. In the case of the Aral Sea crisis, the effects would range from near total depletion of fish stocks, which means the ruin of entire livelihoods and towns, to the possibility of increased rates of cancer, linked to the toxic dusts that blow from the exposed sea bed onto nearby towns. At the end of the first phase, an environmental management assessment is usually planned to examine whether the threat can be resolved by improving the physical infrastructure: building hard structures, like canals, or implementing processes, like waste disposal or bio-remediation.

If the nature and magnitude of the risks are uncertain, the stakeholders see this as a very important issue, and the conditions for resolving the issue are socially and politically complex. Phase two of the process is to consider how to achieve procedural and outcome equity for those affected (strategic and social approaches). Here, the role for scientists involved in the original environmental assessment changes dramatically. When livelihoods are threatened and emotional issues such as increased rates of cancer among children are raised, claims for cool objectivity usually fall on deaf ears. It is at this point that the environmental problem turns into a controversy. Scientists who made the impact assessments will often be asked to take sides, either with the social group that feels they have been wronged, or with the party that the social group feels acted wrongly. Depending on the scale of the controversy, scientists for each side can be found to legitimate the claims and counter claims that are made as the controversy escalates. A dispute is generally entered into by the party – an individual, organization, or national government – that feels wronged, and asks for reparation from the party it perceives to be responsible.

At this point in the risk/controversy cycle, it is crucial that the unit of analysis be at the meso and micro levels of institutions and groups, not the macro level of the nation-state. Perceptions of the risks involved, and concepts of security as they are expressed in social institutional arrangements, need to be elicited. The ethnic, religious and historical intricacies of the region under study need to be clearly understood, and the different views of what would be fair procedure and outcome in mitigating the environmental risk should be identified and classified. In environmental and trade agreements, Western environmental standards, and the perspectives that go with them, have been criticized by developing nations as being imperialistic and unfair. This is true of almost every agreement where a cessation in activities that the West perceives is harmful, is demanded of the developing nations where livelihoods may depend on those same activities.

At the *Third Conference of Parties, UN Framework Convention on Climate Change,* held in Kyoto, Japan in December 1997, developing countries were reluctant to commit to measures to reduce greenhouse gases, when countries like the US had not made equal commitments. One Chinese delegate said, 'In the developed world, only two people ride in a car and yet you want us (the developing world) to give up riding on a bus' (Sims and Passacantando, 1997). This type of disagreement over values and proposed solutions to world environmental problems is likely to be the main source of contention between nations in the future, but that does not mean there have to be wars to solve the problems. Neither does conflict have to emerge when dealing with other, more localized, risks; agreements can be brokered if those guiding the process read the signs for why people feel about a given risk differently, and understand on what priorities their perceptions are based.

When it reaches the attention of policy makers, risk – including environmental risk – is already embedded in a social context, and thus demands frameworks of analysis that evaluate not just the technical, physical problems of the given risk situation, but also an understanding of the objectives, goals and allegiances of rational, social behavior (Douglas, 1985). People's perceptions of risk usually have very little to do with technical probabilities. When threatened, their immediate response is to clamber somehow to safety. Different people feel threatened by different things, and this will affect their judgement regarding what action should be taken. Ingar Palmlund (1992: 206) has written,

> Risk is a code word that alerts society that a change in the social order is being requested. Persons and groups have different attitudes to change in the prevailing order. Within the social order risk issues are used to provide leverage in action to change or to defend the existing pattern. Every reference to risk contains a tacit reference to safety.

What some might regard as a threat to security is not necessarily seen as a threat by others. Collectively, our notions of risky events and what security entails are pluralistic, not consensual. They form an overall complete strategy for avoiding risk and achieving a measure of security, which is not achieved by total agreement over what the threats are and what security means. The nation-state reflects the overall strategy in motion; whether it is a democracy or a dictatorship, it takes a position on defending its borders, values and interests. The nation-state cannot afford ambiguities in its strategies or in its policies, and strong leaders are as highly regarded today as they have been throughout history. The nation-state continues to be a major factor in world politics. But for scientific inquiry into environmental security or risk, the nation-state is an awkward unit of analysis. Because the nation-state includes many types of stakeholders with differing perceptions, the aggregate national statistics hide important information in the sum of its many parts. Policies and politics reduce complex and pluralistic issues to comprehensible soundbites that can be easily understood, repeated and acted upon.

Debates over risk and security – and practically everything else – are more complicated, more drawn out, certainly more pluralistic and more representative of the society involved than nation-state analysis typically suggests. Smaller constituencies include government institutions, private companies, and research and development organizations. They include citizens' action networks, NGO's and local government. At this level, we begin to understand how issues that deal with risk and security are mediated in social institutions by people whose concerns include not only – or even principally – statistical probabilities, but also equity concerns. Long before they become political issues at the nation-state level, the problems that could become threats to environmental security are being debated, coped with (or not coped with) at this meso-scale of institutions. In the case of the Aral Sea, it is the collective farms, *kolkhoz*, that are these meso institutions. They have been responsible for the overfarming and poor irrigation practices, albeit under a strict government quota system that continues from the Soviet era. But their structure also provides protection for the social order under which their members live. If there is concern about changing from cotton to a more sustainable crop, and if irrigation improvements are going to cost more than a *kolkhoz* can afford, these concerns will be mediated through the hierarchy of *kolkhoz* members. If the concerns cannot be mediated through the *kolkhoz* hierarchy, then they will work their way to a resolution in local government, or escalate to the national level. For countries in transition, as well as developed nations, these controversies start off small; stability and regional security depend on shifting these issues to a sustainable resolution before they are given the chance to escalate.

Conclusion

As this chapter has tried to show, there is very little agreement about what the term environmental security means, but there is a discussion underway about what security should be, and how risk analysis helps us understand how to build that security on a pluralistic basis. Security in the post-Cold-War era raises loaded questions about the appropriateness of Cold War assumptions about potential conflicts. In the current trend to call difficult and unpleasant political solutions by softer, more pleasant sounding names, we need to be acutely aware of what actions we call for and on what basis we call for them. Are we really looking for ways to maintain peace and security for all, or inadvertently manufacturing more reasons to go to war? Academics and other researchers bear a responsibility to uncover ways to understand the multiple dimensions of a potential problem. Another burden of responsibility lies with policymakers, first to understand those dimensions, and then work toward stable and peaceful outcomes.

Unfortunately, the current debates over environmental security and the redefinition of security have become mired in the drive to reach some sort of consensus. We argue that that is the worst direction in which to head at this time. Healthy debate and a perusal of every available option are invaluable, if a sophisticated strategy is to be formulated to cope with environmental security problems before they reach the critical level of the nation-state; that is, before military conflict becomes a possibility. Indeed, notions of security, and potential threats to it, should be contended, contradicted and counter-contradicted, if the field is to be strengthened. Premature calls for consensus and for narrowing the debate, run the risk of missing the complexities of the security problems we stand to face in the next millennium.

Risk and security are intertwined. 'Environmental security' refers to the risks to societal stability from pollution or damage to natural resources. One extreme form of this potential for instability might be natural resource wars, but far more likely are increased poverty and famine, deforestation and over-fishing, leakage of radioactive materials and other pollutants. The tools of risk analysis will help to deal with such threats to security. Environmental security problems are, first and foremost, risk problems that must be dealt with at the institutional level, with fairness and pluralism.

Notes

1 For an in-depth analysis of the philosophical and historical meanings of security at the national, international and global levels, and the evolution and development of theory for each term, see Helga Haftendorn, 1991.
2 On March 28, 1979 at Three Mile Island outside Harrisburg, Pennsylvania in the United States, the reactor core at the Three Mile Island (TMI) Unit 2 Nuclear Generating Station was damaged after a minor malfunction

occurred in the system which feeds water to the steam generators. This event led eventually to the most serious commercial nuclear accident in US history and fundamental changes in the way nuclear power plants were operated and regulated.

3. On Saturday, July 10, 1976 an explosion in a chemical plant in Seveso Italy released a chemical mixture in the form of an aerosol cloud into the air in a south-easterly direction. The dioxin filled mixture fell mainly on the communes of Seveso, Meda, Cesano Maderno and Desio. Four days later the first signs of skin inflammation occurred in children and on Sunday, July 25, 1976, fifteen days after the incident, an evacuation of the area was prepared.

4. Thalidomide was developed in the 1950s to prevent nausea during pregnancy. Thalidomide was synthesized in West Germany in 1957 by Chemie Grünenthal and marketed around the world from 1958–1962 in approximately 46 countries under many different brand names. It was not realized that thalidomide could cross the placental wall and affect the foetus. When thalidomide was taken during the first trimester of pregnancy, it caused severe birth defects and premature deaths.

5. On December 3, 1984 a Union Carbide pesticide producing plant leaked a highly toxic cloud of methyl isocyanate onto the densely populated region of Bhopal, central India. Thousands died in the hours and days following the leak. Thousands of survivors in the years following the disaster have been shown to still suffer from one or several of the following ailments: partial or complete blindness, gastrointestinal disorders, impaired immune systems, post-traumatic stress disorders, and menstrual problems in women. A rise in spontaneous abortions, stillbirths, and offspring with genetic defects has been noted.

6. On April 26, 1986 a major reactor accident occurred at the Chernobyl Nuclear Power Plant. There was severe radioactive contamination in the area, resulting in the evacuation of people from a 30-km zone around the power plant. Acute radiation injuries and deaths occurred among plant workers and firemen. Thousands of people involved in the rescue, clean-up and containment operations were exposed to radiation. In the months following the accident, it became clear that radioactive contamination of varying severity had also occurred hundreds of kilometers from the site.

7. In November 1986, a fire at the Sandoz chemical factory in Basel, Switzerland caused a major leak of contaminated water into the Rhine River. The factory had contained 840 tons of pesticides, fungicides, dyes and other toxic chemicals; these were mixed with the large quantities of water that were poured onto the flames. As a result of the accident, 30 tons of highly toxic waste entered the waters of the Rhine.

8. The Kara Kum canal takes water from the Amu Darya and transports it to the Mary oasis and Ashgabat, the capital of Turkmenistan. Over 1000 kilometers in length, the canal ends at Gyzylarbat in western Turkmenistan, not far from the Caspian Sea.

9. Because the NGO movement changes on a regular basis in Central Asia, to find out about NGO development in the region the reader is referred to the Institute for Soviet-American Relations (ISAR) http://

www.isar.org/isar which publishes a quarterly journal, *Surviving Together*, on grassroots cooperation in Eurasia.
10 For an in-depth look at technology choice, irrigation, hydropower, and local environmental sustainability, the reader is referred to research conducted in the Himalayas by Thompson, Warburton and Hatley (1986).

11
Not Seeing the People for the Population: a Cautionary Tale from the Himalaya
Michael Thompson

In the early 1980s, the General Assembly of the United Nations identified a number of patches on the Earth's surface – the Himalaya among them – as environmental 'hot spots': places where, if something was not done, the environment and its human inhabitants would experience catastrophic collapse. UNEP (the United Nations Environment Programme), having succeeded in being given the mandate as the 'lead agency' for these hot-spots, found itself faced with a considerable challenge. It had to arrive at a clear understanding of what the problem in each of these hot-spots was, and it then had to come up with the solutions. UNEP set about this challenge in a way that will be familiar to anyone who has attended a demonstration of some wonderful new piece of military hardware. After a few words from the appropriate high-ranking officers, and to the accompaniment of many 'yessirs' and snappy salutes, responsibility passes rapidly down the line until it reaches the lance-corporal and his trooper side-kick who, as the crew of the death-dealing machine, have to make the thing actually work. In the case of UNEP and the Himalaya, that crew was myself and my research assistant, Michael Warburton (Map 11.1).

My purpose in recounting this snippet of now-quite-ancient history is twofold: first, to suggest that, since much of the recent interest in environmental security seems to be stuck in that 1980s time-warp, we can benefit directly from the lessons that have now been learnt about the interactions between the Himalayan environment and its human inhabitants. Second, to raise a more general warning over what we are getting ourselves into when we link

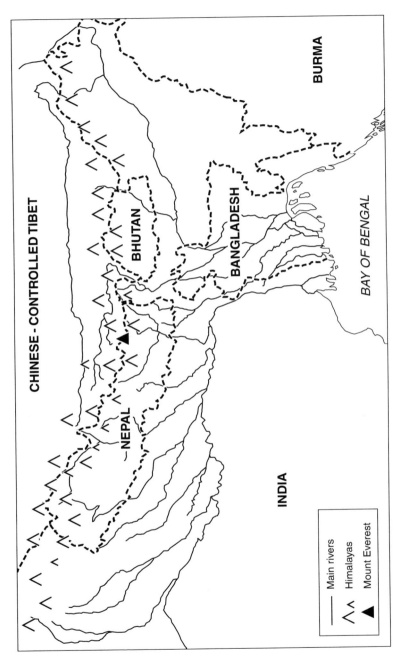

Map 11.1 The Himalaya and their immediate surroundings.

'environment' and 'security'. Put bluntly, if it is the lance-corporals and the troopers who have the environmental expertise, and if security is seen as falling within the realm of 'high politics', then the whole notion of environmental security is going to be deeply unsettling for the security community. If only the lance-corporals and the troopers know what is going on, and if only the top brass can deal in high politics, then perhaps we should see this new interest as a sign that the time has come to put the low-down into the high-up (or vice-versa)!

Actually doing this requires that we reconsider two of the central tenets in present approaches. First, if environmental insecurity is seen as stemming from an *increasing population* combined with a static or, worse still, a declining resource base (eg Renner, 1996; Matthews, 1993; Myers, 1994), then the entire concept will have to be reconsidered if it turns out that population has little, if anything, to do with it. Second, if civil strife, rising international tension, and destabilizing flows of refugees are seen as stemming from the failings of a *weak state* – a state that is unable to take the steps necessary to keep its population from exceeding its territory's carrying capacity (eg Homer-Dixon, 1994) – then there will be no consideration given to the possibility that both environmental degradation and increased instability can stem from the actions of states that are *too strong*.

Incredible though these two possibilities may seem to those who have now framed environmental security in terms of increasing populations and weak states, the lance-corporal's eye-view on the Himalaya suggests otherwise. First, the whole 'downward spiral' argument, with an increasing population that is having to support itself on a declining resource base as its primary vicious circle, is now in tatters. It is the *institutional capabilities* of the people who live in the Himalaya, not their numbers, that determine whether the various crucial spirals are upward or downward. Second, institutional capabilities, at any scale level – household to state – are much influenced by those at the higher and lower levels. Inappropriate interactions, such as the nationalization of Nepal's forests in the 1950s, can severely reduce institutional capabilities at the lower levels by, for example, destroying the village-level commons managing arrangements that regulated forest use; in contrast, appropriate interactions, such as the devolving of forests to the village level, as is now happening in Nepal, combined with the provision of technical expertise from the state-level Forestry Service, can increase dramatically institutional capabilities at all levels.

I will be explaining these serious shifts in the understanding of the problem throughout this chapter, but before I do so, I should point out that there is something suspect, not to say Orwellian, about the way the words 'weak' and 'strong' have been applied to states in relation to environmental security. If the Nepalese state had not been able to guarantee its sovereignty and maintain its borders – a basic, but reasonable enough definition of state strength – then, given the cockpit nature of its geographical position (see Map 11.1, above), it would not exist. Over the past 50 years or so, Nepal has skilfully played off its powerful neighbors – China and India – against one another, and when, more than a century and a half ago, it was in danger of being swallowed up by the then major world power, it inflicted a decisive defeat on the British army. 'Weakness', as used in relation to environmental security, boils down to the state not controlling the size of its population, the implicit assumption being that if it was strong, it could and should do this; both of these are highly questionable propositions. But, either way, state weakness, defined in this way, can have little bearing on environmental insecurity, if population increase is not what is at the bottom of it anyway!

This, however, is not to say that the environment has nothing to do with security – only that there is nothing to be gained, and much to be lost, by thinking in terms of simple (and measurable) 'drivers': population, for instance, and the strength of the state. It is the ingenuity, the skill, the knowledge and the trust that are generated among and between the Nepalese *people*, whose counted heads constitute Nepal's population, that matter. And it is these qualities, appropriately harnessed, nurtured and responded to by the Nepalese state, that promote security, one major part of which comes by way of the environment. People and their institutional capabilities are not unmeasurable, nor is it impossible to devise indicators of appropriate interaction between the various institutional levels, from state to household. But before these things can be done, we have to systematically detach environmental security from its present anchorings in population increase and state strength.

Tragedy in Shangri-La

PRED – Population, Resources, Environment and Development – was the vogue acronym at the time my research assistant and I undertook to provide UNEP with a 'systems overview' of the Himalaya and its problems. The idea, essentially, was that we should identify all the various flows (inputs, outputs, feedbacks and so on) between the people of the region and their physical surroundings. With these identified, and with some quantities placed on them, we would then be able to see just how bad things were. We could also, of course, see what scope there might be for re-arranging some of these flows so as to unlock the disastrous downward spirals and re-configure them into virtuous upward ones: sustainable development, as it is now called.

That this enormous human-cum-physical system *was* locked into all these downward spirals – deforestation, soil erosion, overgrazing, decreasing fertility of the soil that had not yet been eroded, groundwater depletion, the silting-up of reservoirs and the worsening of flooding down in the plains, to mention some of the main ones – had already been established during the 1970s. This was done, most famously, at the Stockholm Environment Conference in 1972 (where UNEP was founded), but also in its various spin-offs and follow-ups: the International Workshop on the Mountain Environment (in Munich, in 1974), for instance, and the book *Losing Ground* (Eckholm, 1975).

In a passage entitled, 'Tragedy in Shangri-La', Eckholm (1975, 764) sets out the key features of the system that we wished to describe in detail sufficient for us to be able to assess what sort of interventions would be needed to avert the looming catastrophe:

> population growth in the context of a traditional agrarian technology in forcing farmers onto ever steeper slopes, slopes unfit for sustained farming even with the astonishingly elaborate terracing practised there. Meanwhile, villagers must roam further and further from their homes to gather fodder and firewood, thus surrounding most villages with a widening circle of denuded hillsides. Ground-holding trees are disappearing fast among the geologically young, jagged foothills of the Himalayas, which are among the most easily erodable anywhere. Landslides that destroy lives, homes, and crops occur more and more frequently throughout the Nepalese hills.

Here then, in stark outline, is precisely the situation that is now identified by many as the root cause of environmental insecurity: an increasing population having to support itself on a resource base that it is actually causing to decline. Grim though it is, this is not the end of Nepal's problems. As the resource base slides away from under its farmers, it causes havoc in the downstream countries – India and Bangladesh – each of which has quite enough problems on its plate already. Eckholm (1975, 764–5) puts it like this:

> Topsoil washing down into India and Bangladesh is now Nepal's most precious export, but one for which it receives no compensation. As fertile soil slips away the productive capacity of the hills declines, even while the demand for food grows inexorably. . . .
>
> The fields now receive less manure than in the past – well below the full amount necessary to preserve high fertility. This is partly because herd sizes have not grown as rapidly as the cultivated area; the hills are already over-grazed and fodder of any kind, whether tree leaves or forage crops, is scarce. Even more ominously, farmers facing an unduly long trek to gather firewood for cooking and warmth have seen no choice but to adopt the self-defeating practice of burning dung for fuel.

Nor is it just themselves that the Nepalis are defeating. As they propel ever more topsoil into their mountain torrents, they render the reservoirs in India useless with startling rapidity, provoke worse flooding in both India and Bangladesh, and aggrade the river-beds to such an extent as to cause 'the river courses to meander about, often destroying prime farmland as they go' (Eckholm, 1975, 765).

All in all, the problems are so daunting, and the spirals so set in their downward ways, that it would seem that nothing can be done. But, back in 1975, there was just one possibility: the opening up of the previously malarial, and therefore unsettled, strip of low-lying territory between the hills of Nepal and the frontier with India: the Terai.

> The presence of undeveloped arable land in the Terai does provide Nepal with some breathing space in which to reverse the downward spiral of population growth, land destruction and declining productivity in which it is now caught. (Eckholm 1975, 765)

But, even here, there was only a small window of opportunity, and even that was being lost thanks to the weakness of the Nepalese state.

The Nepalese government knows that the controlled settlement of the Terai is an essential step to help take some of the pressure off the hills, but it lacks the capacity to keep the land rush under any meaningful control.... If migration down into the Terai continues at the pace of the last 10 years [1965–75], all the good farmland will be occupied in little more than a decade. (Eckholm, 1975, 765)

This, then, was the background to this particular environmental 'hot-spot', and to the 'systems overview' that UNEP had commissioned us to produce. Since this was the early 1980s – already well into the decade of breathing space that Eckholm saw as the region's last chance – there was no time to be lost. We set about our task with an urgency that is uncharacteristic of academic life, gathering data, from the vast literature on the Himalaya, on all the crucial flows, and assembling them into a systems framework: a diagram of boxes and connecting arrows which, when labeled and quantified, would make clear the interlocking nature of all the downward spirals, and provide some guidance on which of them were the most crucial and what would be involved in reversing them.

To our astonishment, we found that uncertainties regarding the crucial quantities were so great that there was no way of telling, in the case of any particular spiral, whether it was upwards or downwards.[1] Deforestation, for instance, which was seen as perhaps the most crucial of all the downward spirals, was understood as resulting from the increasing population using up the forest trees faster than they could grow, and there had been many research projects aimed at estimating these two key rates. Estimates of *per capita fuelwood consumption*, we found, varied by a factor of 67, and estimates of the *sustainable yield from forest production*, which depends on both the rate at which the forest itself grows and the amount of forest there is, varied by a factor of more than a hundred. If you took the two most pessimistic estimates, which is what most people and institutions, wittingly or unwittingly, did, then the Himalaya would be stripped of all vegetation almost overnight. If you took the two most optimistic estimates, which none did, since to do so would be to assert that there was no deforestation problem, then the Himalaya would shortly sink beneath the largest accumulation of biomass the world had ever seen. Much the same held for the other variables we gathered and collated. For instance, it is debatable whether the hill farmers of Nepal are doing *anything* to accelerate the vast process of natural erosion on the lands on which they have made their homes. Their terraced and irrigated fields act as settling beds for enormous volumes of silt-laden water which, in the absence of any human intervention, would have thundered unchecked onto the plains below. And the millions of baskets full of manure that are laboriously carried out to (and, often enough, *up* to) their fields, mean that these farmers are like colonies of rather perverse ants, endlessly carrying their eggs up a 'down' escalator. Indeed, the uncertainties are so pervasive and so overwhelming, that the Theory of Himalayan Environmental Degradation (THED), which is what all these interlocking downward spirals give rise to, is no more plausible than its exact opposite: the Theory of Himalayan Environmental Conservation (THEC).

Both these theories, however, are profoundly flawed in that they assume that there *are* such things as the *per capita fuelwood consumption rate*, and that there *are* processes such as deforestation. There aren't! For there to be a *per capita* fuelwood consumption rate each individual would have to be standardized: an automation, as it were, pre-programmed to consume a fixed amount of wood each day and suffering serious (perhaps terminal) difficulties if unable to lay hands on it. But people are not like that. If fuelwood is plentiful and easily accessible, they will likely use rather a lot; if it becomes scarce and hard to get at, they will likely use less. Sure enough, in Khumbu, the valley below Mount Everest, the Sherpas,[2] in the 1950s, kept their fires going all day and much of the night. More recently, their village forests have deteriorated – as a result of their nationalization, not local population growth[3] – and the Sherpas now only light their fires to cook. They have also been experimenting with solar cookers, electric cookers (powered by their own micro and mini-hydro projects), and new designs of stove that are more fuel-efficient. At the same time, the Nepal state, realizing the inappropriateness of its nationalization policy, has now returned the village forests to local control, thereby much improving their management and eliminating the 'nibble effect'[4] that had so quickly distanced the forests from the villagers.

Teasing out all the details of how these adjustments have happened has required the efforts of specialists from a wide range of disciplines (Thompson, Warburton and Hatley, 1986; Ives and Messerli, 1989; Chapman and Thompson, 1995), but a beginners' course in economics would be enough to tell you to expect them. Yet somehow, the people of Nepal were not granted this minimum of human dignity. Lumped together and reduced to the letter P in PRED, they became indistinguishable from cattle: their needs as fixed as the carrying capacity of the environment on which they had to support themselves. That denial of human dignity, not the imminent collapse that was predicted by all those worthies who gathered in Stockholm, is the tragedy in Shangri-La!

If the people are the problem then they can't be the solution

Embedded within the seemingly scientific, objective and value-free framing that is provided by PRED and by its successor, the I-PAT equation (Ehrlich and Holdren, 1974), is just one model of the person: *the ignorant and fecund peasant*.[5] Essentially indistinguishable from cattle, the hill farmers of Nepal, according to this model, go on and on expanding their population way beyond their environment's carrying capacity, and the state, for its part, is so weak that it cannot pull hard enough on the relevant policy levers: the levers that would stabilize and eventually reverse that runaway population growth.

This was the model of the person that, largely unwittingly, was built into the Stockholm Environment Conference and into its offspring, UNEP. Had it not been, they could not have framed the problem in the way they did. It is also the model that was made explicit and shown to be wholly invalid by all those lance-corporals and troopers who, post-Stockholm, found themselves charged with the task of making it operational. This is not to say that there are no environmental problems in the Himalaya and that everything is wonderfully secure in that part of the world. Nor is it to deny that some people, at certain times and under certain circumstances, do behave in the manner specified by the ignorant and fecund peasant model. But what we *do* now know is that the problem is absolutely not what it was unquestioningly assumed to be. We also know that the peasants of Nepal are not ignorant, nor is their fecundity something that is wholly beyond their control. It is, therefore, deeply depressing to find that, 20 years on, most proponents of environmental security are still wedded to this worthless way of defining the problem. Even more depressingly, they are not alone. Indeed, only among the lance-corporals and the troopers will you find a total rejection of the ignorant and fecund peasant model.[6] Elsewhere, among the top-brass, he is still alive and well.

David Attenborough (1984) – the presenter of the BBC television series 'The Natural World', who, by his own admission, had spent three and a half years 'rushing around the planet on one of the grandest of all possible Grand Tours' – describes the problem as follows:

> According to a recent United Nations Environment Programme survey, one third of the surface of the earth is in danger of turning to desert. . . . We walked across hillsides in Nepal that have been stripped of their trees for firewood. Rain had gouged deep ravines down them, carrying away the soil, and the people were going hungry. A thousand miles away, in the delta of the Ganges, that same soil is being deposited, clogging the river channels. During the rainy season, the water, no longer held back in the forests, rushes down the rivers and floods the delta. Hundreds of people drown and thousands lose their fields and their homes.

If people in Nepal are doing this to the citizens of India and Bangladesh, and that is what this Stockholm/UNEP model tells us they *are* doing, then we have a *causus belli*. But, as the lance-corporals have now shown, this *is not* what is happening. In other words, environmental security, as currently framed, is itself a major cause of insecurity (and of environmental degradation). It is not just a useless concept; it is a positively harmful one. Take, for example, the latest report from Britain's Overseas Development Agency (ODA, 1997, p. 69). This report, which tells us about the Agency's various interventions around the world, trots out the time-honored definition of the problem:

> The Southern Himalayan region of Nepal is facing severe environmental degradation. The growing population's requirement for more food leads to the clearing of forest to provide more land for crop production. Soil becomes exposed and is easily washed away by heavy monsoon rainfall. Land productivity quickly declines, leading to a demand for more land on which to produce crops. In addition there may be an increased risk of floods, reduction of low season water flows, increased erosion and changes in water quality.

This report from the ODA came through my letter-box on the very morning that I received an e-mail (from Ajaya Dixit, in

Kathmandu) about what is going on right in the middle of this southern Himalayan region that the ODA is doing its best to save. The ignorant and fecund peasants of Ramechhap District, far from finding that they can no longer feed themselves from their ever-eroding and increasingly infertile fields, are merrily exporting vast quantities of high quality potatoes to the plains of India. Known as *Godar aloo*, these potatoes currently retail for 65 rupees a *pathi*, while ordinary potatoes retail for 25 rupees. What is more, these ignorant and fecund peasants have been doing this for years – probably at least 100 years – and over that time they have been exporting more and more potatoes. In 1997, they exported a record 20 million rupees-worth. That is about 2000 rupees for every man, woman and child in Ramechhap District: big potatoes by Nepalese standards, and fair-sized ones even by European standards!

What is so extraordinary about this orthodox definition of the problem is that it has David Attenborough standing on the very edge of the abyss that is the self-same abyss that Erik Eckholm was standing on the very edge of ten years earlier. How fortunate for each of them that they should have been there, in Nepal, at just that climactic moment, and how strange that they were standing there ten years apart when the collapse was due to happen within, at the very most, ten years! Similar predictions – '25 years to baldness' (Begley, Moreau and Mazumdar, 1987) – have been made for Himalayan deforestation over the past 30 or so years, yet there are as many, if not more, trees now than there were then: an observation that Bruno Messerli has wryly used to point to a new crisis: that we now have only one or two years left in which to get rid of all these trees! (Messerli, personal communication).[7] Exactly the same apocalypse-postponement has now been demonstrated for the ignorant and fecund peasant model in Africa (Leach and Mearns, 1996) which suggests that students of environmental security should be busy disengaging themselves from this sort of framing, not binding themselves into it.[8]

Returning to our southern hills of Nepal, we should note that, if the ODA's definition of the problem was correct, then what is happening there would be impossible. The potato farmers, their potatoes and their potato fields would have disappeared years ago, leaving a tree-less and soil-less desert in the hills and hectare upon hectare of inundated and silt-plastered Indians in the plains below. What this means, quite simply, is that the ODA is committed to solving a problem that demonstrably cannot be the problem, and the question it raises is: how much more of this sort of development aid can the people of Nepal afford to bear? How much longer, in other words, before security at every level, from household to nation-state, breaks down as a result of the imposition of a solution that has been framed in terms of the ignorant and fecund peasant?

The implications for students of environment and security

'Do no harm' is a pretty sound first principle: one that translates straight into a moratorium on any notion of environmental security that is founded on the conjunction of a growing population, a static or declining resource base, and a 'weak' state. What, then, are we left with? The answer is: the sort of discerning and person-respecting understanding that has been achieved by those – the lance-corporals and troopers – who have been responsible for the demolition of the once-dominant orthodoxy built upon the unreflexive assumption that there is a direct link between population increase and environmental degradation. Everything, rather, hinges on that which is cut out by this assumption: the various patterns of relationships among the various kinds of persons whose counted and undifferentiated heads constitute the P in PRED (and in I-PAT).

Take, for example, all the subtle interactions of forms of social solidarity, ways of knowing and family-level strategies that characterize the lives of our Himalayan villagers. When, as often happens, they set about making themselves a little more secure by engaging in a spot of 'off-farm employment' – portering for tourist treks, or going off on trading expeditions over into Tibet or down into India – they are binding themselves into one well-known form of social solidarity: the *market*. But when they appoint their forest guardians, and respect their judicious stipulations about the year's level of permissible use of the village forest, they are binding themselves into a rather small-scale and face-to-face *hierarchy* (a hierarchy that, as we have seen, came into security-sapping conflict with a larger-scale and rather impersonal hierarchy when the forests were nationalized). And when, as happens with the famous Chipko Movement, they drop everything and rush out to hug the trees (*Chipko* means to hug) to prevent them being appropriated by rapacious logging contractors, aided and abetted by corrupt forestry officers, they are binding themselves into an *egalitarian* form of solidarity. Finally, when these resourceful and social-capital-rich patterns of interaction fail, as one of them did when the village forests were nationalized, then the distinctive morality that underpins each of them is destroyed, the controls that are built into each of them break down, trust ebbs, and purposive action gives way to *fatalism*. Only then do we find people behaving in the way that is consistent with the model of the person that underlies the notion of environmental security.

Security, we begin to see, has to do with the interplay of these different solidarity-induced styles, at each and every scale level – household to nation-state to international regime – and between all those levels. It is those complex dynamics that generate and erode security, and since so many of those solidarity-based transactions involve soil, water, manure, vegetation, livestock and so on, the environment is intimately wrapped up with security. But, if you are going to say anything sensible about environment and security, you are going to have to start with people (a plurality of models of the person), not population (a single model of the person).

There is nothing inherently ignorant or fecund about peasants whose ethno-ecology and accumulated skills enable them to predict where land-slides are likely to occur, and then deliberately trigger them so as to create land that is stable enough to support irrigated rice-fields rather than just the rough grazing that was all that was feasible on the landslip-prone slopes.[9] Nor is there anything weak about a state that, once it has realized the inappropriateness of its policy of nationalization, hands the forest back to the villagers and directs its Forest Service towards strengthening the capabilities of their village-level institutions.

The idea that what is needed is a state strong enough to install, and then pull on, the levers that will bring a whole mass of ignorant and fecund peasants back into line, can never be right. It can never lead to an increase in security but, as this cautionary tale makes clear, it can all too easily result in the opposite.

Notes

1 See chapter entitled, 'Which Way the Spiral?' (pp. 11–31) in Thompson, Warburton and Hatley (1986).
2 Sherpas are people of Tibetan origin who moved into some of the higher and, at that time unsettled, valleys in Nepal (of which Khumbu is one) five or six hundred years ago. World famous because of their mountaineering achievements, they are just one of the vast number of ethnic

groups of which Nepal is comprised. Nepal, since 1990, is a parliamentary democracy with a constitutional monarch.

3 Prior to their nationalization, these forests were managed by village-level commons managing institutions. Nationalization destroyed this face-to-face system of control, replacing it with a remote centralized system that was ineffective and, because of its inability to appreciate and respond to the villagers' perceptions and needs, distrusted (Thompson, Warburton and Hatley, 1986).

4 Without effective institutional arrangements, people overuse the most accessible forest and underuse its further reaches. The result is a progressive retreat of forest edge, even when the overall extraction rate is no more than it was when the forest was being managed effectively.

5 The I-PAT equation, which holds that environmental impact (I) is some multiplication of population (P), affluence (A) and technology (T), still pins the blame firmly on increasing population. It then compounds the environmental impact by insisting that the more prosperous people become, the more natural resources they will use, and by further insisting that technological development, by making it easier and cheaper for the people to get all these material things, will make things even worse.

The 'ignorant and fecund peasant' (first introduced in Ives and Messerli, 1989) is a convenient shorthand description of the social construction (or model) of the person that is built into these ways of framing the problem. Those who frame the problem this way do not, of course, see themselves as ignorant and fecund peasants. Their own model of the person, in other words, is very different, which means that the environmental effects of population will depend not on its numerical strength but on the relative proportions of these persons within it. Nor, of course, is it safe to assume that there are only two models. For a critical treatment of these serious shortcomings of the orthodox approach, see Thompson (1998).

6 Their key findings are nicely summarized in Cohen (1988).

7 The same reasoning, expressed in a more scholarly way, is set out in Messerli and Hofer (1995).

8 Why all these institutions and individuals zero-in so unerringly on this particular way of framing the problem, and on this particular model of the person, is the question we need to answer if we are to do anything sensible about development and about environmental security. The answer, at bottom, is that our convictions about how the world is and how people are, are part-and-parcel of our *social solidarities*: the various ways in which we bind ourselves to one another. This is the province of those social scientists who take serious account of the social construction of reality. Unfortunately, there is not the space to explain this approach here, but it is set out in general terms in Thompson and Rayner (1998), in terms of the Himalaya and its environmental problems in Thompson (1995), and in relation to security and the environment in Thompson (1997 and 1998). A vignette is included in the concluding paragraphs of this chapter.

9 Painstaking anthropological research (Johnson, Olson and Manandhar 1982; Kienholtz *et al.*, 1984) has revealed that this is what is actually

going on in the Middle Hills of Nepal, thereby overturning the orthodox view about land-slides, and land degradation generally, which was based on the assumption that these ignorant and fecund peasants have no understanding of the geomorphology of the landscape they are interacting with. It is the steady accumulation of these sorts of research efforts, often but not exclusively in the Himalaya, that has now transformed our understanding of the interaction of human and natural systems to the point where what was seen as the solution – a 'strong' state – is now very much part of the problem (Batterbury, Forsyth and Thompson, 1997).

Bibliography

Daily Report: East Asia.
Far Eastern Economic Review.
Japan Times Weekly (International Edition).
The Korea Herald.

Abruzere, Robin L. 1998. 'Middle East Realignment Spreads Into the Caucasus'. *Global Intelligence Updates: Red Alert*. Austin, Tex.: STRATFOR Systems, Inc. Internet: http://www.stratfor.com. March 12.

Agnew, J. 1998. *Geopolitics: Revisioning World Politics*. London: Routledge.

Akan, Burcu and Lauren Van Metre. 1997. 'Dayton Implementation: The Return of Refugees'. *Special Report*. Washington, DC: United States Institute of Peace.

Akiner, S. 1993. 'Environmental Degradation in Central Asia' from Panel VI *Economics of the Environment*, NATO Economic Colloquium Economic Developments in Cooperation Partner Countries from a Sectoral Perspective. July.

Aladin, Nikolai. 1995. 'The Conservation of Ecology of the Podonidae from the Caspian and Aral Seas'. *Hydrobiologia*. 307: 85–97.

Alker, H.R. and P.M. Haas. 1993. 'The Rise of Global Ecopolitics' in *Global Accord: Environmental Challenges and International Responses*. Ed. Nazli Choucri. Cambridge, MA: MIT Press, 133–71.

Alley, R.B. and M.L. Bender. 1998. 'Greenland Ice Cores: Frozen in Time'. *Scientific American*. 278(2) February: 80–5.

Allison, R. ed. 1996. *Challenges for the Former Soviet South*. Brookings Institute and The Royal Institute of International Affairs.

Aryal, Manisha. 1995. 'Dams: The Vocabulary of Protest'. *Himalayan Magazine*. July–August.

Attenborough, David. 1984. 'Threat to the Living Planet'. *The Observer*. London, April 1.

Bächler, Gunther. 1994. 'Desertification and Conflict: The Marginalization of Poverty and of Environmental Conflict'. *Occasional Paper of the Environment and Conflicts Project (ENCOP)*. Zurich: Center for Security Studies and Conflict Research, No. 10.

Banfield, E.C. 1958. *The Moral Basis of a Backward Society*. Free Press, Glencoe.

Batterbury, Simon; Timothy Forsyth; and Koy Thompson. 1997. 'Environmental Transformations in Developing Countries: Hybrid Research and Democratic Policy'. *The Geographical Journal*. 163.2: 126–32.

Beaumont, Peter. 1991. 'Transboundary Water Disputes in the Middle East'. Presented at a conference on Transboundary Waters in the Middle East, Ankara, September.

Bedeski, Robert E. 1995. 'Unconventional Security and the Republic of Korea: A Preliminary Assessment'. Cancaps Paper. No. 8, August.

Begley, Sharon; Ron Moreau; and Sudip Mazumdar. 1987. 'Trashing the Himalayas'. *Newsweek*. November 9.

Bennett, Bruce. 1996. 'Implications of Proliferation of New Weapons on Regional Security'. 11th Conference on Korea–U.S. Security Studies on 'The Search for Peace and Security in Northeast Asia Toward the 21st Century'. Seoul, Korea. October 25–25. Santa Monica. Calif.: RAND Corporation.

Bennett J.W. and K.A. Dahlberg 1990. 'Institutions, Social Organization, and Cultural Values' in *The Earth as Transformed by Human Action*. Ed. B.L. Turner *et al.*, New York: Cambridge University Press.

Bernstein, B. 1971. 'On the Classification and Framing of Educational Knowledge' in *Class, Codes, and Control* Vol. 1. London: Routledge and Kegan Paul.

Biswas, A.K. and T. Hashimoto, eds, 1996. *Asian International Waters: From Ganges–Brahmaputra to Mekong*. Oxford: Oxford University Press.

Botkin, D. 1990. *Dischordant Harmonies: a New Ecology for the Twenty First Century*. Oxford: Oxford University Press.

Bradbury, J. 1993. 'Risk Communication in Environmental Restoration Programs'. Paper presented for Battelle Pacific Northwest Laboratories at the Interagency Conference 'The Risk Assessment Paradigm After Ten Years: Policy and Practice, Then, Now, and the Future'. Wright-Patterson Air Force Base, April 5–8.

Bradbury, Judith; Kristi M. Branch; Judith Heerwagen; and Edward B. Liebow. 1994. *Community Viewpoints of the Chemical Stockpile Disposal*. Washington, DC: Battelle.

Braudel, F. 1981–84. *Civilization and Capitalism, 15th–18th Century*. New York: Harper and Row.

Brecher, Michael and Jonathan Wilkenfeld. 1997. *A Study of Crisis*. Ann Arbor, Michigan: University of Michigan Press.

Brock, L. 1997. 'The Environment and Security: Conceptual and Theoretical Issues' in *Conflict and the Environment*. Ed. N.P. Gleditsch. Dordrecht: Kluwer Academic Publishers, 17–34.

Brown, Lester. 1977. 'Redefining Security'. *Worldwatch Paper No. 14*. Washington, DC: Worldwatch Institute.

Bryant, R.L. and S. Bailey. 1997. *Third World Political Ecology*. London: Routledge.

Bullock, John and Adel Darwish. 1993. *Water Wars: Coming Conflicts in the Middle East*. London: St. Dedmundsbury Press.

Butts, Kent. 1994. 'Why the Military Is Good for the Environment' in *Green Security or Militarized Environment*. Ed. Jyrki Kakonen. Brookfield: Dartmouth Publishing Company.

—— 1997. 'The Strategic Importance of Water'. *Parameters*. Spring: 65–83.

Buzan, Barry. 1991. *People, States and Fear: the National Security Problem in International Relations*. Boulder, Colorado: Lynne Rienner Publishers.

Buzan, B.; O. Wæver and J. de Wilde. 1998. *Security: A New Framework for Analysis*. Boulder: Lynne Rienner.

'Central Europe Survey'. 1995. *Economist*. (November 18): 5.

Cernea, Michael. 1993. 'The Sociologist's Approach to Sustainable Development'. *Finance and Development* 30:4, 11–13.

Chapman, Graham P. and Michael Thompson, eds, 1995. *Water and The Quest for Sustainable Development in The Ganges Valley*. London: Mansell.

Chern, Wen S. and William S. James. 1988. 'Measurements of Energy Productivity in Asian Countries'. *Energy Policy* 16 October: 503.

Clausewitz, Carl Von. 1832. *On War*. Princeton, NJ: Princeton University Press. Reprint: 1976.
Cohen, Nick. 1988. 'Doomsday View of Floods Nonsense: Scientists Say'. *The Independent*. London. (September 17).
Conca, Ken. 1994. 'Rethinking the Ecology-Sovereignty Debate'. *Millennium*. 23(3). 701–11.
—— 1994. 'In the Name of Sustainability: Peace Studies and Environmental Discourse'. *Peace and Change*, vol. 19 (2): 91–113.
Connelly, Matthew and Paul Kennedy. 1004. 'Must It Be the West Against the Rest?' *The Atlantic Monthly* 274: 61–83.
Cooley, John. 1984. 'The War Over Water'. *Foreign Policy*. No. 54 Spring: 3–26.
Cooper, Jerrold. 1983. *Reconstructing History from Ancient Inscriptions: The Lagash–Umma Border Conflict*. Malibu, Calif.: Undena.
Crosby, A. 1986. *Ecological Imperialism: The Biological Expansion of Europe 900–1900*. Cambridge: Cambridge University Press.
'Curbing a Pollution Economy'. 1991. *South*, 119, February: 20.
Dabelko, Geoffrey D. 1996. 'Ideas and the Evolution of Environmental Security Conceptions'. ISA Paper.
Dabelko, Geoff and P.J. Simmons. 1997. 'Environment and Security: Core Ideas and U.S. Government Initiatives'. *SAIS Review* 17. Winter/Spring: 127–146.
Dalby, Simon. 1992. 'Security, Modernity, Ecology: the Dilemmas of Post-Cold War Security Discourse'. *Alternatives*, vol. 17, 1.
—— 1992. 'Ecopolitical Discourse: Environmental Security and Political Geography'. *Progress In Human Geography* 16, 4, 503–22.
—— 1995. 'Neo-Malthusianism in Contemporary Geopolitical Discourse: Kaplan, Kennedy and New Global Threats'. ISA Paper.
—— 1996. 'The Environment as Geopolitical Threat: Reading Robert Kaplan's Coming Anarchy'. *Ecumene* 3, 4. 472–96.
—— 1998. 'Geopolitics and Global Security: Culture, Identity and the "Pogo Syndrome"'. in *Rethinking Geopolitics*. Eds G. Ó Tuathail and S. Dalby London: Routledge. 295–313.
—— 1998b. 'Ecological Metaphors of Security: World Politics in the Biosphere'. *Alternatives*. 23, 3. 291-319
—— 1999. 'The Threat from the South: Geopolitics, Equity and Environmental Security' in *Contested Grounds: Security and Conflict in the New Environmental Politics*. Eds D. Deudney and R. Matthew Albany: State University of New York Press.
Dauvergne, P. 1997. *Shadows in the Forest: Japan and the Politics of Timber in South East Asia*. Cambridge, MA: MIT Press.
Davis, Uri, Antonia Maks, John Richardson. 1980. 'Israel's Water Policies'. *Journal of Palestine Studies*. 9, 2 Winter: 3–32.
Deibert, Ronald J. 1996. 'Military Monitoring of the Environment'. *Environmental Change and Security Project Report* 2: 28–32.
Del Rosso, Stephen J. 1995. 'The Insecure State: Reflections on "the State" and 'Security' in a Changing World'. *Daedalus* 124:2. Spring.
Dellapenna, Joseph. 1994. 'Treaties as Instruments for Managing Internationally-Shared Water Resources: Restricted Sovereignty vs. Community of Property'. *Case Western Reserve Journal of International Law*, 26.

—— 1995. 'Building International Water Management Institutions: the Role of Treaties and Other Legal Arrangements'. in *Water in the Middle East: Legal, Political, and Commercial Implications.* Eds J.A. Allan and C. Mallat. London and New York: Tauris Academic Studies: 55–89.

De Soto, Hernando. 1989. *The Other Path: The Invisible Revolution in the Third World.* NY: Harper & Row.

Deudney, Daniel. 1990. 'The Case Against Linking Environmental Degradation and National Security'. *Millennium: Journal of International Studies* 19: 461–76.

—— 1991. 'Environment and Security: Muddled Thinking'. *Bulletin of Atomic Scientists.* April: 23–8.

—— 1997. 'The Limits of Environmental Security' in *Flashpoints in Environmental Policymaking.* Eds Sheldon Kamieniecki, George Gonzales and Robert O. Vos. Albany: State University of New York Press. 281–310.

Deudney, Daniel and Richard A. Matthew, Eds, 1999. *Contested Ground:Security and Conflict in the New Environmental Politics.* Albany: State University of New York Press.

Deutsch, Karl W. and Singer, J. David. 1969. 'Multipolar Power Systems and International Stability' in *International Politics and Foreign Policy.* Ed. James N. Rosenau. New York: Free Press.

Diamond, J. 1997. *Guns, Germs and Steel: the Fates of Human Societies.* New York: Norton.

Diehl, P.F., Ed. 1998. Special Issue on Environmental Conflict, *Journal of Peace Research.* 35, 3.

Dillman, Jeffrey. 1989. 'Water Rights in the Occupied Territories'. *Journal of Palestine Studies.* 19, 1. Autumn: 46–71.

Dokken, Karen and Nina Graeger. 1995. 'The Concept of Environmental Security – Political Slogan or Analytical Tool?' *(PRIO), Report 2.* International Peace Research Institute, Oslo.

Douglas, Mary. 1982. *In the Active Voice.* London: Routledge and Kegan Paul.

Douglas, M. 1985. *Risk Acceptability According to the Social Sciences.* New York: Russell Sage Foundation.

Durkheim, E. 1893. *De la Division du Travail Social: Étude sur L'Organization des Sociétés Superieurs.* Paris, Alcan.

Eaton, David and Joseph Eaton. 1994. 'Joint Management of Aquifers between the Jordan River Basin and the Mediterranean Sea by Israelis and Palestinians: an International Perspective'. in 'Proceedings of Joint Management of Shared Aquifers: First Workshop'. Eds Eran Feitelson and Marwan Haddad. Jerusalem June 27–9: 131–52.

Eberstadt, Nicholas and Judith Banister. 1992. 'Divided Korea: Demographic and Socioeconomic Issues for Reunification'. *Population and Development Review* 18, 3: 505–31.

Elliot, L. 1998. *The Politics of the Global Environment.* New York: New York University Press.

Elliott, Michael. 1991. 'Water Wars'. *Geographical Magazine.* May.

Ellsworth, Robert F. 1997. 'America's Security Interests for the New Century'. *Bulletin.* Washington, DC: Atlantic Council of the United States.

Erlich, Paul. 1968. *The Population Bomb.* Ballantine Press.

Esman, M. and N. Uphoff. 1984. *Local Organizations.* Ithaca, NY: Cornell University Press.

Evans, Peter. 1995. *Embedded Autonomy: States and Industrial Transformation*. Princeton: Princeton University Press.
Falkenmark, Malin. 1986. 'Fresh Waters as a Factor in Strategic Policy and Action' in *Global Resources and International Conflict: Environmental Factors in Strategic Policy and Action*. Ed. A.H. Westing. New York, NY: Oxford University Press. 85–113.
Feitelson, Eran and Marwan Haddad. 1995. *Joint Management of Shared Aquifers: Final Report*. A Cooperative research project of the Palestine Consultancy Group, East Jerusalem, and the Harry S. Truman Research Institute, Hebrew University of Jerusalem. December.
Feshbach, Murray. 1995. *Ecological Disaster: Cleaning Up the Hidden Legacy of the Soviet Regime*. New York: Twentieth Century Fund Press.
Finger, M. 1991. 'The Military, the Nation State and the Environment'. *The Ecologist*. 21, 5. 220–5.
Fischer, Dietrich. 1993. 'Non-Military Aspects of Security: a Systems Approach'. *UNIDIR*. Dartmouth: Aldershot.
Fredkin, Phillip. 1981. *A River No More*. New York: Knopf.
Frost, Robert. 1962. *The Road Not Taken*. New York: Holt.
Fukuyama, Francis. 1995. *Trust*. London: Penguin Books.
—— 1996. *Trust: The Social Virtues and the Creation of Prosperity*. New York: Free Press.
Gaddis, John Lewis. 1987. *The Long Peace: Inquiries Into the History of the Cold War*. New York: Oxford University Press.
Gadgil, M. and R. Guha. 1995. *Ecology and Equity: The Use and Abuse of Nature in Contemporary India*. London: Routledge.
Galambos, Judit. 1992. 'Conflict in the Use and Management of International Commons' in *Perspectives on Environmental Conflict and International Relations*. Ed. Jyrki Kakonen. London: Pinter Publishers.
Gertz, Bill. 1996. 'N. Korea Targets U.S. GIs'. *Washington Times*. October 22.
Gizewski, Peter and Thomas Homer-Dixon. 1996. 'Environmental Scarcity and Violent Conflict: the Case of Pakistan'. *The Project on Environment, Population and Security, Peace and Conflict Studies*. University of Toronto.
Gleditsch, N.P., Ed., 1997. *Conflict and the Environment*. Dordrecht: Kluwer Academic Publishers.
—— 1997. 'Environmental Conflict and the Democratic Peace' in *Conflict and the Environment*. Ed. Nils Petter Gleditsch. Dordrecht: Kluwer Academic Publishers, 91–106.
—— 1998. 'Armed Conflict and the Environment: a Critique of the Literature'. *Journal of Peace Research*. 35, 3, 381–400.
Gleick, Peter. 1991. 'Environment and Security: The Clear Connections'. *Bulletin of the Atomic Scientists*. April: 17–21.
—— 1993. 'Water and Conflict: Fresh Water Resources and International Security'. *International Security*, 18.
Goshko, John M. 1996. '9 Million Still Uprooted By Soviet Union's Demise'. *Washington Post*, (April 28).
Græger, Nina. 1996. 'Environmental Security?' *Journal of Peace Research* 33, 1: 109–16.
Gyawali, D. and A. Dixit. 1994. 'The Himalaya Ganga: Contending with Interlinkages in a Complex System' in *Water Nepal* 4.1: 1–6.

Haftendorn, Helga. 1991. 'The Security Puzzle: Theory-Building and Discipline-Building in International Security'. *International Studies Quarterly* No. 35: 3–17.
Hammer, W. 1980. *Product Safety Management and Engineering*. NJ: Prentice-Hall.
Hamner, Jesse H. and Aaron T. Wolf. 1998. 'Patterns in International Water Resource Treaties: the Transboundary Freshwater Dispute Database'. *Colorado Journal of International Environmental Law and Policy*. 1997 Yearbook.
Hardin, Garrett. 1968. 'The Tragedy of the Commons'. *Science*. 162: 1243–8.
Hayton, Robert D. 1982. 'The Law of International Aquifers'. *Natural Resources Journal*. 22, 1 January: 71–94.
—— 1988. *River And Lake Basin Development*. New York: UN Department of Technical Co-Operation and Development, Natural Resources Water Series, No. 20.
—— 1991. 'Reflections on the Estuarian Zone'. 31 *National Resources Journal*.
Hayton, Robert and Albert Utton. 1989. 'Transboundary Groundwaters: the Bellagio Draft Treaty'. *Natural Resources Journal*, 29, Summer.
Healey, Derek and Wilfred Lutkenhorst. 1989. 'Export Processing Zones: the Case of the ROK', *Industry and Development* 26: 7: 1–56.
Hecht, S, and A. Cockburn. 1990. *The Fate of the Forest: Developers, Destroyers and Defenders of the Amazon*. Harmondsworth: Penguin.
Heller, Mark and S. Nusseibah. 1991. *No Trumpets, No Drums: a Two-State Solution of the Israeli-Palestinian Conflict*. New York: Hill and Wang.
Hillel, Daniel. 1994. *Rivers of Eden: the Struggle for Water and the Quest for Peace in the Middle East*. New York: Oxford University Press.
Hirschman, Albert O. 1971. *A Bias for Hope*. Yale University Press.
—— 1982. *A Dissenter's Confession: the Strategy of Economic Development Revisited*. Pioneer Lectures. IBRD/World Bank.
—— 1984. 'Against Parsimony'. *American Economic Review Proceeding* 74:93.
Holl, Jane E. Executive Director. 1996. *Carnegie Commission on Deadly Conflict*. Washington, DC: Carnegie Corporation of New York.
Homer-Dixon, Thomas. 1990. 'Environmental Change and Violent Conflict: Understanding the Causal Links', paper prepared for the workshop 'Environmental Change and Threats to Security'. *American Academy of Arts and Sciences*. (March 30–31).
—— 1991. 'On the Threshold: Environmental Changes as Causes of Acute Conflict'. *International Security*, vol. 16: 76–116.
—— 1993. 'Environmental Change and Violent Conflict'. *Scientific American*, February: 38–45.
—— 1994. 'Environmental Scarcities and Violent Conflict: Evidence from Cases'. *International Security* vol. 19(1): 5–40.
—— 1995. 'The Ingenuity Gap: Can Poor Countries Adapt to Resource Scarcity?' *Population and Development Review* 21:3. September 587–612.
—— 1996. 'Debate'. *Woodrow Wilson Center Environmental Change and Security Project Report* Issue 2. Spring: 49–57.
—— 1996. 'Environmental Scarcity, Mass Violence, and the Limits to Ingenuity'. *Current History*. November. 359–65.
Homer-Dixon, Thomas and Marc Levy. 1995/96. 'Correspondence'. *International Security* 20. Winter: 189–98.

Homer-Dixon, Thomas and Valerie Percival. 1996. *Environmental Scarcity and Violent Conflict: Briefing Book.* Toronto: University of Toronto Press.

Hong, Seoung-Yong. 1991. 'Assessment of Coastal Zone Issues in the ROK'. *Coastal Management: an International Journal of Marine Environment.* Resources, Law and Society 19: 391–415.

Hopkirk, Peter. 1994. *The Great Game: the Struggle for Empire in Central Asia.* Kodansha.

Housen-Couriel, Deborah. 1994. 'Some Examples of Cooperation in the Management and Use of International Water Resources'. Hebrew University of Jerusalem, Truman Research Institute for the Advancement of Peace, March.

Howitt, R., J, Connell and P. Hirsch, Eds, 1996. *Resources, Nations and Indigenous Peoples.* Melbourne: Oxford University Press.

Huntington, Samuel P. 1996. *The Clash of Civilizations and the Remaking of the World Order.* New York: Simon & Shuster.

Hyman, A. 1996. 'Post-Soviet Central Asia' in *Challenges for the Former Soviet South.* Ed. R. Allison. Brookings Institute and The Royal Institute of International Affairs.

Ignatieff, Michael. 1993. *Blood and Belonging: Journeys into the New Nationalism.* New York: Noonday Press.

Intergovernmental Panel on Climate Change. 1992. *Climate Change: The IPCC Scientific Assessment.* Cambridge: Cambridge University Press.

Ives, Jack D. and Bruno Messerli. 1989. *The Himalayan Dilemma: Reconciling Development and Conservation.* London and New York: Routeledge/United Nations University.

Johnson, K.; E.A. Olson and S. Manandhar. 1982. 'Environmental Knowledge and Response to Natural Hazards in Mountainous Nepal'. *Mountain Research and Development.* 2.2: 175–88.

Johnston, B.R., ed. 1994. *Who Pays the Price? The Sociocultural Context of Environmental Crisis.* Washington: Island Press.

Käkönen, J. 1988. *Natural Resources and Conflicts in the Changing International System.* Brookfield, NY: Avery Press.

Kang, Peter C. 1988. 'Political and Corporate Group Interests in South Korea's Political Economy'. *Asian Profile* 16:3: 210–11.

Kaplan, Robert. 1994. 'The Coming Anarchy'. *Atlantic Monthly.* 273, 2. Spring: 44–73.

Kaye, Lincoln. 1989. 'The Wasted Waters'. *Far Eastern Economic Review.* (Feb 2).

Keller, Kenneth. 1996. *Environmentalism and Security.* MIT Security Studies Program. Cambridge, Mass.: Massachusetts Institute of Technology.

Kennedy, Paul and Matthew Connelly. 1994. 'Must it be the West against the Rest?' *Atlantic Monthly.* 274: 61–83.

Keohane, R. and M. Levy, Eds, 1996. *Institutions for Environmental Aid Cambridge.* MA: MIT Press.

Keohane, Robert O. and Joseph Nye Jr. 1977. *Power and Interdependence.* Boston: Little, Brown.

Kienholtz, H.; G. Schneider; M. Bichsel; M. Grunder; and P. Mool. 1984. 'Mapping of Mountain Hazards and Slope Stability'. *Mountain Research and Development.* 4,3: 247–66.

Kiho, Young Whan and Dong Suh Bark. 1981. 'Food Policies in a Rapidly

Developing Countries: The Case of South Korea'. *The Journal of Developing Areas* 16, 1. October: 47–70.

Kim, Hyung Kook. 1982. 'Social Factors From Rural to Urban Areas With Special Reference to Developing Countries: the Case of Korea'. *Social Indicators* 10: 29–74.

Kim, Sang-Beom. 1997. 'For the Record'. *East Asian Review*. Winter.

Kliot, Nurit, Deborah Shmueli, and Uri Shamir. 1997. *Institutional Frameworks for the Management of Transboundary Water Resources*. Haifa, Israel: Water Research Institute (two volumes).

Klotzli, S. 1994. 'The Water and Soil Crisis in Central Asia – a Source for Future Conflicts?' *Occasional Paper of the Environment and Conflicts Project (ENCOP)*. Zurich: Center for Security Policy and Conflict Research, No. 11.

Kolars, J.F. and W.F. Mitchell. 1991. *The Euphrates River and the Southeast Anatolia Development Project*. Carbondale, IL.: Southern Illinois University Press.

Krause K. and M. Williams, Eds, 1997. *Critical Security Studies: Concepts and Cases*. Minneapolis: University of Minnesota Press.

Lang, Winfried. 'Negotiation in the Face of the Future' in 'Negotiation and Global Security: New Approaches to Contemporary Issues'. *American Behavioral Scientists*. Ed. Celia Albin. Special Edition Vol. 38/No. 6.

Lasswell, Harold D. 1950. *National Security and Individual Freedom*. New York: McGraw-Hill.

Leach, Melissa and Robin Mearns, Eds, 1996. *The Lie of the Land: Challenging Received Wisdom on the African Environment*. Oxford and Portsmouth, New Hampshire: The International African Institute in Association with James Currey and Heinemann.

Levy, Marc. 1995a. 'Time for a Third Wave of Environment and Security Scholarship?' *The Environmental Change and Security Project Report*. Washington, DC: Woodrow Wilson International Center for Scholars, 44–6.

—— 1995b. 'Is the Environment a National Security Issue?' *International Security*. Vol. 20(2): 35–62.

Lewis, M.W. and K.E. Wigen. 1997. *The Myth of Continents: A Critique of Metageography*. Berkeley: University of California Press.

Libiszewski, Stephan. 1992. 'What is an Environmental Conflict?' *Occasional Paper of the Environment and Conflicts Project (ENCOP)*. Zurich: Center for Security Studies and Conflict Research, No. 1.

—— 1995. *Water Disputes in the Jordan Basin Region and their Role in the Resolution of the Arab–Israeli Conflict*. Zurich: Center for Security Studies and Conflict Research, Occasional Paper No. 13. August.

Lilienthal, David. 1951. 'Another Korea in the Making?' *Collier's Magazine* (4 April).

Lindblom, C. 1977. *Politics and Markets: the World's Political and Economic Systems*. New York: Basic Books.

Lipschutz, Ronnie D. 1989. *When Nations Clash: Raw Materials, Ideology and Foreign Policy*. New York: Ballinger.

—— 1997. 'Environmental Conflict and Environmental Determinism: the Relative Importance of Social and Natural Factors' in *Conflict and the Environment*. Ed. N.P. Gleditsch. Dordrecht: Kluwer Academic Publishers, 35–50.

—— 1997. *Global Civil Society and Global Environmental Governance: The Politics of Nature from Place to Planet*. Albany: State University of New York Press.
Lonergan, Stephen. 1993. 'Impoverishment, Population and Environmental Degradation: the Case for Equity'. *Environmental Conservation*. Vol. 20(4): 328–34.
—— 1997. 'Linking Environment and Security' in Presentation to the 2nd Open Science Meeting of the Global Environmental Change Community, Vienna. June.
Lonergan, Stephen and David Brooks. 1994. *Watershed: the Role of Freshwater in the Israeli–Palestinian Conflict*. Ottawa: IDRC Books.
Lovelock, J.E. 1988. *The Ages of Gaia: a Biography of Our Living Earth*. New York: Norton.
Lowi, Miriam R. 1992. 'West Bank Water Resources and the Resolution of Conflict in the Middle East'. Occasional Paper Series of the Project on Environmental Change and Acute Conflict. (1 September). University of Toronto and the American Academy of Arts and Sciences.
—— 1993. [2nd edition, 1995]. *Water and Power: the Politics of a Scarce Resource in the Jordan River Basin*. Cambridge, UK: Cambridge University Press.
—— 1993b. 'Bridging the Divide: Transboundary Resource Disputes and the Case of West Bank Water'. *International Security* 18, 1. Summer: 113–38.
—— 1995. 'Rivers of Conflict, Rivers of Peace'. *Journal of International Affairs* 49, 1, Summer.
—— 1996. 'The Hydrogeography of the Middle East at Peace: Israel/Palestine Water Issues'. paper delivered at the conference, *Water: A Trigger for Conflict/A Reason for Cooperation*. Indiana Center on Global Change and World Peace. Indiana University (Bloomington). 7–10 March.
—— 1996b. 'Political and Institutional Responses to Transboundary Water Disputes in the Middle East'. *Environmental Change and Security Project Report*. 2. Spring. The Woodrow Wilson Center.
—— 1999. 'Transboundary Resource Disputes and their Resolution', in Daniel Deudney and Richard Matthew. Eds, *Contested Grounds: Security and Conflict in the New Environmental Politics*. Albany: State University of New York Press.
—— 1999b. 'Water and Conflict in the Middle East and South Asia: Are Environmental and Security Issues Linked?' *Journal of Environment and Development*, 8: 4, December.
Maine, H.S. 1861. *Ancient Law*. London: J. Murray.
Malcolm, Noel. 1998. 'A Political Flashpoint Left In Limbo By Outside World'. *Daily Telegraph*, March 4.
Mandel, Robert. 1992. 'Sources of International River Basin Disputes'. *Conflict Quarterly*. 12, 4. Fall: 25–56.
Mannion, A.M. 1997. *Global Environmental Change: A Natural and Cultural Environmental History*. London: Addison Wesley Longman.
Mardon, Russell. 1990. 'The State and the Effective Control of Foreign Capital: The Case of South Korea'. *World Politics* 43: 111–38.
Matthew, Richard A. 1995. 'Environmental Security and Conflict: an Overview of the Current Debate'. *National Security Studies Quarterly* 1: 1–10.
—— 1995. 'Environmental Security: Demystifying the Concept, Clarifying the Debate'. *Environmental Change and Security Project Report* 1: 14–23.

―― 1996. 'The Greening of American Foreign Policy'. *Issues in Science and Technology* XII:1, 39–47.
―― 1997. 'Rethinking Environmental Security' in *Conflict and the Environment*. Ed. N.P. Gleditsch. Dordrecht: Kluwer Press: 71–90.
―― 1998. 'Security and Scarcity: a Common Pool Resource Perspective' in *Common Pool Resources and the Environment*. Eds S. Barkin and G. Shambaugh. Albany: SUNY Press.
Mathews, Jessica Tuchman. 1989. 'Redefining Security'. *Foreign Affairs* 68. Spring: 162–77.
―― 1994. 'The Environment and International Security', in Michael T. Klare and Daniel C. Thomas, Eds. *World Security*. New York. St Martin's Press.
―― 1996. 'Overlooking the "POPs" Problem'. *The Washington Post*. March 11.
McCaffrey, Stephen C. 1993. 'The Evolution of the Law of International Watercourses'. *Australian Journal of Public & International Law*. 45: 87.
McKibben, Bill. 1989. *The End of Nature*. New York: Random House.
Messerli, Bruno and Thomas Hofer. 1995. 'Assessing the Impact of Anthropogenic Land Use Change in the Himalayas' in *Water and the Quest for Sustainable Development in the Ganges Valley*. Eds Graham P. Chapman and Michael Thompson. London: Mansell. 64–89.
Midlarsky, M.I. 1998. 'Democracy and the Environment: an Empirical Assessment' *Journal of Peace Research*. 35(3). 341–61.
Mill, J.S. 1861. *Utilitarianism*. London.
Millas-Estany, Esperanza. 1998. Catalanist. Daughter of the Chief-of-Staff to the President of the Republic of Catalunya (1936–39). Personal Interview. McLean, Va.: March 30.
Mische, P.M. 1989. 'Ecological Security in an Inter-Dependent World'. *Breakthrough*. Summer/Fall: 7–17.
Mohammed, Nadir A.L. 1994. 'The Development Trap: Militarization, Environmental Degradation and Poverty and Prospects of Military Conversion'. Organization for Social Science Research in Eastern Africa. Occasional Paper 5.
―― 1996. 'Environmental Conflicts in Africa'. *NATO Advanced Research Workshop* Paper.
Muldavin, J.S.S. 1997. 'Environmental Degradation in Heilongjiang: Policy Reform and Agrarian Dynamics in China's New Hybrid Economy'. *Annals of the Association of American Geographers*. 87(4). 579–613.
Mumford, L. 1934. *Technics and Civilization*. New York: Harcourt, Brace and Co.
Myers, Norman. 1986. 'The Environmental Dimension to Security Issues'. *The Environmentalist*. 6, 4: 251–7.
―― 1989. 'Environment and Security'. *Foreign Policy* 74. Spring: 23–41.
―― 1994. *Ultimate Security: the Environmental Basis of Political Stability*. New York: Norton.
Naff, Tom and Ruth Matson, Eds, 1984. *Water and Peace in the Middle East*. Amsterdam: Elsevier.
Narain, Sunita. 1997. 'Environment and Inequality' in Presentation to the 2nd Open Science Meeting of the Global Environmental Change Community, Vienna. June.
Nelkin D., and M. Pollack. 1980. 'Consensus and Conflict Resolution: the Politics of Assessing Risk' in Dierkes et al. *Technological Risk: Its Perception and Handling in the European Community*. Massachusetts: 65–75.

ODA (Overseas Development Agency, London). 1997. *Natural Resources Research: Working for Development*. (Third report on the ODA's Renewable Natural Resources Research Strategy).

Oren, I. and J. Hays. 1997. 'Democracies Rarely Fight One Another, But Developed Socialist States Rarely Fight at All'. *Alternatives*. 22(4). 493–521.

O'Riordan, T; C.L. Cooper; A. Jordan; S. Rayner; K.R. Richards; P. Runci; S. Yoffe; 1998. 'Institutional Frameworks for Political Action' in *Global Choice and Climate Change, Volume, The Societal Framework*. Eds S. Rayner and E.L. Malone. Columbus: Batelle Press.

Orr, Robert M., Jr. 1987. 'The Rising Sun: Japan's Foreign Aid to ASEAN, the Pacific Basin, and the ROK'. *Journal of International Affairs* 41: 39–62.

Ó Tuathail, G. 1996. *Critical Geopolitics: The Politics of Writing Global Space*. Minneapolis: University of Minnesota Press.

Palmlund, I. 1992. 'Social Drama and Risk Evaluation' in *Social Theories of Risk*. Eds S. Krimsky and D. Golding. Westport, CT: Praeger. 197–214.

Paterson, M. 1996. *Global Warming and Global Politics*. London: Routledge.

Perelet, Renat. 1994. 'The Environment as a Security Issue' in *Environment: Towards a Sustainable Future*. Amsterdam: Kluwer, 147–73.

Perry, Duncan. 1998. 'Destiny on Hold: Macedonia and the Dangers of Ethnic Discord'. *Current History*, March.

Perry, William J. 1996. 'Remarks at the John F. Kennedy School of Government, Harvard University'. *News Release* 278–96. Washington, DC: Department of Defense.

Ponting, Clive. 1991. *A Green History of the World: the Environment and the Collapse of Great Civilizations*. New York: St. Martin's Press.

Postel, Sandra. 1997. *Last Oasis: Facing Water Scarcity*. Worldwatch Institute.

Putnam, Robert D. 1993. *Making Democracy Work: Civic Traditions in Modern Italy*. Princeton, NJ: Princeton University Press.

Ramet, Sabrina P. 1998. 'The Slovenia Success Story'. *Current History*, March.

Ratzel, F. 1897. *Politische Geographie*. Munich: Oldenbourg.

Rayner Steve. 1987. 'Learning from the Blind Man and the Elephant, or Seeing Things Whole in Risk Management' in *Uncertainty in Risk Management, Risk Assessment and Decision Making*. Eds Vincent Covello, Lester B. Lave, Alan Moghissi and V.R.R. Uppuluri. New York: Plenum Press, 207–12.

—— 1995. 'A Conceptual Map of Human Values' in *Equity and Social Considerations Related to Climate Change*. Ed. A. Katama Nairobi: ICIPE Science Press.

Rayner, Steve and R. Cantor. 1987. 'How Fair is Safe Enough? The Cultural Approach to Societal Technology Choice' in *Risk Analysis* Vol. 7 No. 1 (1): 3–13.

Redclift, M. 1996. *Wasted: Counting the Costs of Global Consumption*. London: Earthscan.

Remans, Wilfried. 1995. 'Water and War'. *Humantäres Völkerrecht* 8, 1.

Renn, Ortwin. 1992. 'Concepts of Risk' in *Social Theories of Risk*. Eds Sheldon Krimsky and Dominic Golding. Westport, CT: Praeger.

Renner, Michael. 1989. 'National Security: the Economic and Environmental Dimensions'. *Worldwatch Paper No. 89*. Washington, DC: Worldwatch Institute.

—— 1996. *Fighting for Survival: Environmental Decline, Social Conflict, and the*

New Age of Insecurity. New York and London: Norton for Worldwatch Institute.

Richards, Alan. 1983. 'Economic Imperatives and Political Systems'. *Middle East Journal*. 47, 2 (spring): 217–27.

—— 1992. 'Food, Jobs, and Water: Governance and Participation for a Sustainable Agriculture'. Paper prepared for: *Alexandria International Conference; Sustainability of Egyptian Agriculture in the 1990s and Beyond*, May 15–19.

Robinson, John B. 1992. 'Risks, Predictions and Other Optical Illusions: Rethinking the Use of Science in Social Decision Making'. *Policy Sciences* 25: 237–54.

Rogers, Peter. 1994. 'The Agenda for the Next Thirty Years', in Peter Rogers and Peter Lydon. Eds *Water in the Arab World: Perspectives and Prognoses*. Cambridge: Harvard University Press.

Romm, Joseph J. 1993. *Defining National Security: the Non-Military Aspects*. New York: Council on Foreign Relations Press.

Rowe, W.D. 1977. *An Anatomy of Risk*. New York: Wiley.

Rummel, R. J. 1997. *Power Kills: Democracy as a Method of Nonviolence*. New Brunswick: Transaction Press.

Russett, B. 1993. *Grasping the Democratic Peace: Principles for a Post-Cold War World*. Princeton NJ: Princeton University Press.

Sachs, A. 1995. *Eco-Justice: Linking Human Rights and the Environment*. Washington, Worldwatch Institute, Worldwatch Paper 127.

Samson, Paul and Bertrand Charrier. 1997. 'International Freshwater Conflict: Issues and Prevention Strategies'. Green Cross Draft Report, May.

Sandholtz, Wayne; Michel Borrus; John Zysman; Ken Conca; Stowsky; Steven Vogel; Steve Weber. 1992. *The Highest Stakes: the Economic Foundations of the Next Security System*. Oxford University Press.

Schmida, Leslie. 1983. *Keys to Control: Israel's Pursuit of Arab Water Resources*. Washington, DC: American Educational Trust.

Schmidt, Fabian. 1998. 'Upheaval in Albania'. *Current History*, March.

Serageldin, Ismail. 1995. 'Earth Faces Water Crisis'. Washington, DC: World Bank Press Release (August 6).

Sessions, George, ed. 1995. *Deep Ecology for the Twenty-First Century*. New York: Random House.

Shapiro, M. 1997. *Violent Cartographies: Mapping Cultures of War*. Minneapolis: University of Minnesota Press.

Shaw, Brian. 1996. 'When Are Environmental Issues Security Issues?' *Woodrow Wilson Center Environmental Change and Security Project Report*. Issue 2, Spring: 39–44.

Simon, Julian. 1981. *The Ultimate Resource*. Princeton. NJ: Princeton University Press.

—— 1989. 'Lebensraum: Paradoxically, Population Growth May Eventually End Wars'. *Journal of Conflict Resolution* 33: 164–80.

Sims, Kelly and John Passacantando. 1997. *Ozone Action Report*. from: Third Conference of Parties UN Framework Convention on Climate Change. Kyoto, Japan (December 15).

Smil, V. 1993. *Global Ecology: Environmental Change and Social Flexibility*. London: Routledge.

Smith, A. 1759. *The Theory of Moral Sentiments*. London: A. Millar.

Smith, N. 1990. *Uneven Development: Nature, Capital and the Production of Space.* Oxford: Blackwell.
Snyder, Scott. 1997. *North Korea's Decline and China's Strategic Dilemmas.* Washington, DC: United States Institute of Peace.
Sorenson, T.C. 1990. 'Rethinking National Security'. *Foreign Affairs* 69. 1–18.
Soroos, Marvin. 1992. 'Conflict in the Use and Management of International Commons' in *Perspectives on Environmental Conflict and International Relations.* Ed. Jyrki Kakonen. London: Pinter Publishers.
—— 1994. 'Global Change, Environmental Security and the Prisoner's Dilemma'. *Journal of Peace Research* Vol. 31, 3: 317–32.
—— 1997. *The Endangered Atmosphere: Preserving a Global Commons.* Columbia: University Of South Carolina Press.
Spolar, Christine. 1998. 'Macedonia Maintains Calm Despite Violence in Nearby Kosovo'. *Washington Post* (March 16).
Springer, Allen L. 1996. 'Unilateral Action in Defense of Environmental Interests: an Assessment'. ISA Paper.
Spykman, Nicholas John. 1942. *America's Strategy in World Politics: the United States and the Balance of Power.* New York: Archon Books. Reprint: 1970.
Starr, Joyce. 1991. 'Water Wars'. *Foreign Policy* No. 82, Spring.
Stauffer, Thomas. 1982. 'The Price of Peace: the Spoils of War'. *American–Arab Affairs*, 1, Summer.
Stewart, C. 1997. 'Old Wine in Recycled Bottles'. Paper for the annual meeting of the British International Studies Association, Leeds, December.
Stoever, William A. 1986. 'Foreign Investment as an Aid in Moving From Least Developed to Newly Industrializing'. *The Journal of Developing Areas* 20: 223–47.
Stork, Joe. 1983. 'Water and Israel's Occupation Strategy'. *MERIP Reports.* 116, 13 July–August: 19–24.
Teclaff, Ludvik A. 1991. 'Fiat or Custom: the Checkered Development of International Water Law'. *Natural Resources Journal*, 31.
Thompson, Michael. 1995. 'Policy-making in the Face of Uncertainty: the Himalaya as Unknowns' in *Water and the Quest for Sustainable Development in the Ganges Valley.* Eds Graham P. Chapman and Michael Thompson. London: Mansell. 25–38.
—— 1997. 'Security and Solidarity: an Anti-reductionist Framework for Thinking About the Relationship Between Us and the Rest of Nature'. *The Geographical Journal.* 163.2: 141–9.
—— 1998. 'The New World Disorder: Is Environmental Security the Cure?' *Mountain Research and Development.* 18.2: 117–22.
Thompson, M. and D. Gyawali. 1998. 'Transboundary Risk Management in the South: a Comparative Example from the Himalaya'. Working Paper of the Norwegian Center for Research in Organization and Management (LOS-Senteret), Bergen, Norway, LOS-Senter Notat 9721.
Thompson, Michael and Steve Rayner. 1998. 'Cultural Discourses' in *Human Choice and Climate Change.* Eds Rayner, Steve and Elizabeth L. Malone. Columbus, Ohio: Battelle Press. 265–343, Vol. 1.
Thompson M.; M. Warburton and T. Hatley. 1986. *Uncertainty on a Himalayan Scale.* London: Ethnographica.
Toennies, F. 1887. *Gemeinschaft und Gesellschaft.* Wissenschaftlich, Darmstadt.

Turner, B.L., et al. 1990. *The Earth as Transformed by Human Action*. Cambridge: Cambridge University Press.
Ullman, Richard. 1983. 'Redefining Security'. *International Security*. 8, 1 Summer: 129-53.
UNEP. 1997. *Global Environmental Outlook*. Oxford: Oxford University Press.
United Nations Food and Agriculture Organization. 1978. *Systematic Index of International Water Resources Treaties, Declarations, Acts and Cases, by Basin: Volume I*. Legislative Study No. 15.
—— 1984. *Systematic Index of International Water Resources Treaties, Declarations, Acts and Cases, by Basin: Volume II*. Legislative Study No. 34.
US Department of State. 1995. *Bosnia Fact Sheet: The Road to the Dayton Peace Agreement*. (December 6).
—— 1998. *Bosnia and Herzegovina Country Report on Human Rights Practices for 1997*. (January 30).
US Office of the Secretary of Defense. 1997. *Proliferation: Threat and Response*. Washington, DC: Government Printing Office.
Utton, Albert. 1982. 'The Development of International Groundwater Law'. *Natural Resources Journal*. 22, 1. January: 95.
Verdnadsky V.I. 1945. 'The Biosphere and the Noosphere'. *American Scientist*. 33. 1-12.
Vesilind, Paul. 1993. 'Water: the Middle East's Critical Resource'. *National Geographic* Vol. 183: 5.
Vlachos, Evan. 1990. 'Prologue'. *Water International*. 15, 4. December: 185-8.
Wæver, O. 1995. 'Securitization and Desecuritization' in *On Security*. Ed. Ronnie Lipschutz. New York: Columbia University Press. 46-86.
Walker R.B.J. 1993. *Inside/Outside: International Relations as Political Theory*. Cambridge: Cambridge University Press.
Waltz, Kenneth N. 1969. 'International Structure, National Force, and the Balance of World Power' in *International Politics and Foreign Policy*. Ed. James N. Rosenau. New York: Free Press.
Wapner, P. 1996. *Environmental Activism and World Civic Politics*. Albany: State University of New York Press.
Watts, M. 1998. 'Nature as Artifice and Artifact' in *Remaking Reality: Nature at the Millennium*. Eds Bruce Braun and Noel Castree, London: Routledge. 243-68.
Weiss, Charles, Jr. 1995. *The Marshall Plan: Lessons for U.S. Assistance to Central and Eastern Europe and the Former Soviet Union*. Washington, DC: Atlantic Council of the United States.
Wescoat, James L. Jr. 1995. 'Main Currents in Early Multilateral Water Treaties: a Historical-Geographic Perspective, 1648-1948'. *Colorado Journal of International Environmental Law & Policy* 7, 1: 39-74.
Westing, A.H., ed. 1986. *Global Resources and International Conflict: Environmental Factors in Strategic Policy and Action*. New York, NY: Oxford University Press.
—— 1986. 'Environmental Factors in Strategic Policy and Action: An Overview' in *Global Resources and International Conflict: Environmental Factors in Strategic Policy and Action*. Ed. A.H. Westing. New York: Oxford University Press. 1-20.
—— 1989. 'The Environmental Component of Comprehensive Security'. *Bulletin of Peace Proposals* 20: 129-34.

Williams, Daniel. 1998. 'In Armenia, the Disappointment of Freedom'. *Washington Post*, March 4.

Williamson, O. 1975. *Markets and Hierarchies: Analysis and Antitrust Implications*. New York: Free Press.

Willard, Daniel E. and Avram G.B. Primack. 1996. 'Water a Source of Sustainable Conflict'. Indiana University: School of Public and Environmental Affairs. February.

Wilson, E.O. 1992. *The Diversity of Life*. Cambridge: Harvard University Press.

Wolf, Aaron T. 1995. *Hydropolitics along the Jordan River: Scarce Water and its Impact on the Arab–Israeli Conflict*. Tokyo: United Nations University Press.

—— 1997. 'International Water Conflict Resolution: Lessons from Comparative Analysis'. *International Journal of Water Resources Development*. 13, 3. December.

—— 1999a. 'Criteria for Equitable Allocations: the Heart of International Water Conflict'. *Natural Resources Forum*, 23, 1, February: 3–30.

—— 1999b. 'Water Wars and Water Reality: Conflict and Cooperation along International Waterways' in S. Lonergan. Ed. *Environmental Change, Adaptation, and Security*. Dordrecht, The Netherlands: Kluwer Academic Press.

Wolf, E. 1982. *Europe and the People without History*. Berkeley: University of California Press.

'World Bank Report'. 1997. Staff Appraisal Support: Turkmenistan Water Supply and Sanitation Project No. 16142-TM (April 24).

World Commission on Environment and Development (WCED). 1987. *Our Common Future*. Oxford: Oxford University Press.

World Press Review. 1995. Cover Story: 'Next, Wars Over Water?' November: 8–13.

Xhudo, Guz. 1997. 'What Brought Anarchy to Albania'. *Jane's Intelligence Review*, June.

Yergin, Daniel. 1977. *Shattered Peace: the Origins of the Cold War and the National Security State*. Boston: Houghton Mifflin.

Zartman, I. William. 1991. 'The Structure of Negotiation' in Victor Kremenyuk. Ed. *International Negotiation: Analysis, Approaches, Issues*. San Francisco: Jossey-Bass: 65–77.

Index

Accords/Agreements/Treaties
  Anti-Ballistic Missile 141
  Bellagio Draft 141
  Boundary Waters (US and Canada) 140
  Conventional Armed Forces in Europe 25
  Dayton 22, 23
  Indus Waters 126, 157
  Intermediate Range Nuclear Forces 25
  Jordan–Israel Peace 141
  Nile (1929) 141
  Nile (1959) 138, 141
  Nuclear Non-Proliferation 25
  Palestinian–Israeli 141
  Strategic Arms Reduction 25
  Taba Interim (Oslo II) 80
  transboundary water 124, 128–9, 130
  Vuoska (1972) 141
Afghanistan 184
Albania 19, 20, 21
al-Nasir, Abd 161
Arab League 127, 160
Aral Sea 7, 129, 138, 143, 144, 172, 179, 180, 181, 182, 183, 185, 188
Armenia 18, 19, 29
arms control 25, 26, 31, 105, 121
associational network 51
Aristobulus 180
Aswan High Dam 140
Australia 43
Azerbaijan 18, 19, 29

Babylonia 175
Bangladesh 123, 128, 197, 201
Bekovich, Alexander 180
Berlin 14
Bhopal 175
biological weapons 17
Bosnia 21, 22, 23
Burundi 154

Canada 52, 111, 119, 140
Caspian Sea 180
Caucasus Mountains 180
Central Intelligence Agency (CIA) 42, 45
chemical weapons 17, 176
Chernobyl 24, 176
China 14, 28, 108, 111, 118, 144, 195
Chipko movement 202
Christopher, Warren 33, 67
civic engagement 54
civil society 5, 57, 64, 93, 97, 183
clientism 54, 58
Code of Hammurabi 175
Colombia 119
conflict, interstate 42, 43, 44, 87, 88, 126, 159
conflict, intrastate 43, 72, 85, 87, 126
Croatia 23
Czech Republic 19

Dead Sea 160
deforestation 24, 44, 87, 189, 196, 198, 199, 202
Democratic Republic of Congo 154

Easter Island 34
ecological history 90, 94
ecologists 36, 38
economic security 110
Egypt 127, 141, 142, 153–4, 155, 161
enforcement 142
environment, the 49–64
environmental issues

management 185
risk 172, 175, 176, 185, 186, 187, 189
security 1, 2, 33, 50, 64, 76, 84, 97, 151, 172, 185, 189, 192, 193, 195, 200
Eritrea 142
Eshkol, Levi 161
Ethiopia 127, 154
European Union 18

famine 17, 56, 189
Finland 141
food security 116

Gandak Power Project 140
Gaza 78, 152
Germany 163
Godar aloo 202
Golan Heights 78, 125, 162
Goodman, Sherri 43
Gore, Albert 44
governance 50, 51
Great Britain 139, 141, 142
Great Lakes 129
Greece 18, 21, 29
Green Line 158
Greenpeace 2
groundwater 141, 158
  depletion 196

Haiti 24
Hee, Park Chung 108, 111
Herodotus 180
Himalaya 8, 192, 193, 194, 198, 200, 201, 202
Hindu Kush Mountains 179
Hong Kong 111
Huleh Basin 127
human rights 117
human security 69, 70, 72, 74, 75
humanists 36–7
Hungary 19
hydraulic imperative 125, 162
hydropower 140, 143

India 21, 124, 126, 128, 140, 145, 157, 163, 195, 197, 201, 202
Instability 15, 26, 27, 33, 42, 124

Intergovernmental Panel on Climate Change 92
International Chemical Weapons Convention 176
International Human Dimensions Project (IHDP) 43
International Monetary Fund (IMF) 121
International Red Cross 27
International Workshop on the Mountain Environment 196
Iraq 113, 127, 155, 156, 163, 165
Israel 18, 21, 39, 77, 78, 125, 127, 144, 145, 158, 160, 161, 162, 163, 165
Italy 52, 53, 57, 120

Japan 28, 87, 108, 119, 163, 186,
Jonglei Canal Project 140
Jordan, Kingdom of 78, 144, 156, 160
Jung, Kim Dae 108, 119

Kara Kum canal 179
Kazakhstan 179, 180
Kenya 154
Kolkhoz 185, 188
Korea, North 17, 107–22
Korea, South 6, 103, 107–22
Korean Economic Development Organization (KEDO) 121
Korean War 14, 17, 107
Kuwait 113
Kyrgyzstan 179

Lake Chad 138
Lebanon 78, 160, 161, 162
Lesotho Highlands Project 129

Macedonia 21, 22
Marshall Plan 13, 31
Marshall, George 11, 13
Mauritania 127
MEDEA 44
Mesopotamia 175
Mexico, Mayan 34
Mexico, modern 144

monitoring 139
multifunctionalism 60

Nagorno–Karabakh 18
National Water Carrier
  System 158, 160
Nepal 140, 194–202
Nigeria 86
North Atlantic Treaty Organization
  (NATO) 2, 12, 14, 18, 43, 45
North Pacific Cooperative Security
  Dialogue 105
Northern Ireland 52, 55

Oil 29, 39, 86, 113, 115
  pipelines 19
  spills 113
  tankers 113
Old Testament 175
overgrazing 196
Overseas Development
  Agency 201, 202

Pacific Northwest National
  Laboratory 43
Pakistan 21, 126, 156, 157, 158,
  163
Palestinian population 158, 165
Palestinian territory 158
Pamir Mountains 180
Perry, William J. 3, 11, 16
Peru 119
Peter the Great 180
Poland 19
Portugal 29
poverty 56, 66, 71, 110, 111, 166
preventive defense 3, 11, 16, 24,
  26, 28, 43

radioactive waste 184, 189
refugees 22, 23
regionalism 71
resource security 114
risk assessment 176
  psychometric 176, 177
  social 176, 178
  strategic 176, 177
  technical 176, 177
rivers

Amu Dar'ya (Oxus) 144, 179,
  180
Banias 160, 162
Cauvery 128
Cenpa 125
Colorado 128, 144, 146
Columbia 146
Dan 160
Danube 129, 146
Euphrates 127, 129, 143, 145,
  146, 152, 155, 159, 163, 164
Ganges 123, 128, 129, 130, 143,
  144, 145, 146, 201
Gash 142
Hasbani 160, 162
Indus 124, 126, 129, 130, 143,
  144, 146, 152, 153, 156, 159
Jordan 39, 77, 78, 124, 125,
  127, 129, 143, 144, 152, 153,
  156, 159, 160, 161, 162, 163,
  164, 165
LaPlata 129, 143, 146
Litani 162
Mekong 124, 129, 144
Niagara 140
Nile 124, 127, 129, 140, 142,
  143, 144, 146, 152, 164
Nile, Blue 154
Nile, White 154
Rio Grande 146
Salween 129
Senegal 127, 146
Shatt al-Arab 146
Syr Dar'ya 144, 179, 180
Volga 182
Yarmouk 160, 162
Russia 25, 28, 44, 108, 182
Rwanda 154

Sam, Kim Young 108
Sandoz 175
scarcity 33, 42, 43, 115, 166
  of resource 40, 84
  of water 34, 123, 151, 152
Sea of Galilee (Lake Tiberias) 160
Senegal 127
Serbia 21, 23
Seveso 175
Singapore 111

Slovenia 19, 29
social capacity 57, 73
social solidarity 202
societal resilience 50
solidarity
  egalitarian 61, 62
  hierarchical 61, 62
  market 61, 62
  social 60
Somalia 39, 127
Southeast Anatolia Development Project (GAP) 155, 156
Sovereignty 15, 98, 103, 105, 109, 111, 117
Soviet Union 1, 12–14, 34, 89, 105, 111, 146, 149, 173
Spain 30
stability 7, 12, 13, 14, 15, 24, 31, 50, 96, 123, 185
Statehood 15
statists 37
Stockholm Environmental Conference 196, 200
Straits of Tiran 161
Sudan 127, 141, 142, 154
sustainable development 73, 74, 75, 82, 91, 196
Sweden 44
Switzerland 52
Syria 78, 125, 127, 145, 155, 156, 160, 161, 163, 165

Taiwan 110
Tajikistan 179
Tanzania 154
Terai 197, 198
thalidomide 175
Theory of Himalayan Environmental Conservation (THEC) 199
Theory of Himalayan Environmental Degradation (THED) 199, 200, 202
Third Conference of Parties, UN Framework Convention on Climate Change 186
Threat 25, 35–6, 47, 67, 106, 121, 149, 187
  definition of 1, 3, 15, 106
Three Mile Island 175
Tien Shan Mountains 179
Turkey 18, 29, 145, 155, 165
Turkmenistan 179, 183

United Nations 2, 18, 44, 45, 93, 142
United Nations Conference on Environment and Development (UNCED) 2, 70
United Nations Development Programme (UNDP) 2, 69
United Nations Environment Programme (UNEP) 2, 42, 180, 192, 196, 198, 200, 201
United Nations High Commissioner for Refugees (UNHCR) 23
United States 11–14, 28, 35, 43, 110, 119, 140, 161, 176
United States Government 26–7
US Agency for International Development (USAID) 180
US Army 176
US Army War College 45
Uzbekistan 179, 180

Vietnam 144

Warsaw Pact 14
water
  allocation of 139
  law 129
  wars 7, 123, 124, 126, 147, 160
watershed 123
West Bank 78, 125, 129, 152, 158
Woo, Roh Tae 119
World Bank 157, 160, 180, 183
World Commission on Environmental Development 91
World Conference on Human Rights 118
World War I 11
World War II 12, 14, 28
World Wildlife Fund 2

Yugoslavia, the former 52